AF881089

SOTHIS PRESS

Sothis Press aims to completely reimagine the public domain book with an emphasis on universal access to captivatingly designed print editions. We believe that technology will enable every single book ever published to be perpetually accessible in beautiful print form. Sothis is beauty in ubiquity.

www.sothis.press
1 Waterhouse Square, London, ECIN 2ST

Cement and How to Use It

A Working Manual of Up-to-Date-Practice in the Manufacture and Testing of Cement; The Proportioning, Mixing, and Depositing of Concrete, and Its Application to All Forms of Concrete Construction, Plain and Reinforced; With Special Chapters on Concreting

William A. Radford

SOTHIS PRESS

ISBN 978-1-0368-0770-2

Cement and How to Use It
is in the public domain.

This edition is published by
 The company
is registered in England and Wales.
The Company Number is 14900389.

Cement

AND

How to Use It

A Working Manual of

UP-TO-DATE PRACTICE IN THE MANUFACTURE AND TESTING OF CEMENT; THE PROPORTIONING, MIXING, AND DEPOSITING OF CONCRETE, AND ITS APPPLICATION TO ALL FORMS OF CONCRETE CONSTRUCTION, PLAIN AND REINFORCED; WITH SPECIAL CHAPTERS ON CONCRETING TOOLS AND MACHINERY, WATER-PROOFING, WORKING RULES, ETC.

Edited under the Supervision of

WILLIAM A. RADFORD

PRESIDENT OF THE RADFORD ARCHITECTURAL COMPANY
EDITOR-IN-CHIEF OF "RADFORD'S CYCLOPEDIA OF CONSTRUCTION," THE "AMERICAN CARPENTER AND BUILDER," AND THE "CEMENT WORLD"

ILLUSTRATED

THE RADFORD ARCHITECTURAL COMPANY
CHICAGO, ILL.

PREFACE

IN recent years a profound and widespread popular interest has been aroused in the subjeet of Cement and its application to all forms of Concrete Construction, Plain and Reinforced. As a structural agent, concrete has, in fact, assumed a position of supreme importance, as evidenced by the innumerable instances of its use to be found on every hand, in the construction of buildings of every size and grade, bridges, sidewalks, ornamental forms, etc.

The present volume has been prepared in response to numerous requests for a treatise written in simple, non-technical English, which will serve the purposes of a plain, practical working guide to the selection of cement and aggregates, the principles of proportioning and mixing concrete, and the rules of its intelligent use in the various forms of construction to which it has shown itself so well adapted.

Throughout this volume constant emphasis is laid on the practical as distinguished from the theoretical mode of treatment. The use of higher mathematics and engineering formulae has been avoided, and all phases of the subject reduced to terms of the simplest and clearest English. Practical working rules and labor-saving tables, embodying the useful information that engineers have gathered at great labor, are given in abundance, and all important details explained by clear and carefully executed diagrams. Special chapters are included on the Tools and Equipment necessary for up-to-date work, and on the important subject of Waterproofing; and particular attention is paid to the use of concrete in the erection of houses of block, monolithic, and stucco construction, and the application of concrete to useful purposes around the home and on the farm.

On account of the intimate relation which Cement and Concrete now bear to every branch of construction, the present volume will be of great practical value not only to concrete workers, but to all other craftsmen engaged in the various trades of the Building Industry, such as Carpenters, Steel Workers, Bridge Builders, Plumbers, Sanitary Engineers, and others.

CEMENT

AND HOW TO USE IT

INTRODUCTION

CEMENT is the basis of the most important development affecting the modern structural field. Even in this age of marvelous mechanical progress and engineering achievement, the growth of the modern cement manufacturing industry, and the rapidly extending use of concrete as a material of construction adapted to an endless variety of work, are subjects of world-wide wonder. All branches of the Building Industry have been affected—and some of them have been revolutionized—through the advances made by this new structural agent.

When, not many years ago, the steel-frame structure supplanted the historic masonry and timber, it worked a far-reaching transformation in building methods and materials, which was soon reflected in the external appearance as well as internal arrangement and details of new structures of almost every class. The sky-lines of cities and towns assumed new forms; and especially in the larger industrial centers, the very face of things was changed.

With the advent of a true Portland cement concrete—the handmaid of steel—still another and greater era of structural and industrial evolution was ushered in. Nothing in all the wonderful annals of American industrial development has ever equaled the expansion that has characterized the Portland cement industry during the period since the year 1898. There are in fact many who, in view of the marvelous growth of cement manufacturing in America during these opening years of the twentieth century, and the wonderful versatility of this new structural agent, as shown by innumerable examples of its application to all kinds and grades of construction, are already proclaiming that the **Age of Cement** is upon us. Just as in the manufacturing world the successive eras of power development can be traced through the stages of human labor, water power, and steam power, to the present dawning age of electrical power, so, in the building world—we are told—the claims of wood and stone and brick and tile and iron and steel have now to adjust themselves to the urgent demands for recognition of concrete as a structural factor which is not only capable of supplementing them to advantage, but in many respects far surpasses them all.

In any country subject to rigorous climatic conditions such as prevail in North America, with its excessive changes of temperature and humidity, no natural building material exposed to the elements will withstand the ravages of time. Wood, being organic, is peculiarly responsive to the disintegrating activity of natural agencies, and, unless carefully protected from exposure thereto, is very rapid in its decomposition. Under the atmospheric influences of this climate, all grades of sandstone will in time crumble back to mother earth; marble is still less enduring; and even the so-called "everlasting" granite is similarly limited in its life. Steel is subject to oxidation or rust from the combined action of carbon dioxide and water or the direct action of atmospheric oxygen, and to still more rapid chemical corrosion from the action of sulphurous and other noxious acid gases and liquids; and it must therefore be given the closest attention in the application of protective paints or other coverings to insure any degree of permanence.

Among building materials, concrete stands out alone as the one which can truly be called "permanent," and which has at the same time the quality of progressive and accumulative strength. It is this characteristic of accumulative strength—a strength increasing with the lapse of time—which more than anything else gives concrete its advantage over other materials in engineering and building construction. Under conditions of exposure that in the case of other building materials would accelerate the natural tendencies to decay and final collapse, concrete displays higher and higher powers of resistance, the action of the elements only adding hardness and strength as time goes on.

In addition, however, to the marvelous durability of concrete under trying conditions—which is, perhaps, its most striking feature—no other material now used in construction can be said to meet so successfully the combined requirements of versatility in useful application, adaptability to varying structural conditions and combinations, simplicity combined with dignity and stability of appearance, harmonious adjustment to varying environment, cleanliness and hygienic value, and—last, but not least—economy of cost.

In spite of early unfortunate experiences with natural cement, and more recent instances of actual failure—which have been traceable in every case to ignorance or criminal carelessness—concrete has irresistibly forged its way to general acceptance; has triumphantly risen above prejudice and doubt; and, by its own intrinsic merits, has finally won for itself an assured place as an instrumentality of twentieth century progress whose possibilities when realized may dwarf the dreams of the wildest imagination. Stronger and more durable than any natural stone, unaffected by fire or moisture, capable of adaptation to any position or condition, workable by unskilled labor, lending itself easily to any form of ornamentation, vermin-proof, cleanly, and comparatively inexpensive, it ranks among the foremost of the valuable gifts to mankind from the treasure-house of modern scientific and technical research.

A widespread popular knowledge of the nature and possibilities of concrete, the principles of its composition, and the rules of its intelligent use, will react beneficially in a thousand ways upon the life of the community. It will immeasurably increase the facilities of industry and commerce; it will minister to social conveniences and the education of public and private taste; it will make better homes, better municipalities, and a more progressive citizenship.

THE STORY OF CONCRETE

No one knows when, where, or by whom concrete was invented or first employed. Its use as a material of construction dates back to the dim antiquity of prehistoric times, and examples of the ancient craft have come down to us with practically undiminished strength which have withstood the bombardment of nature's forces through all the centuries. Orpheus himself, whose lyric strains animated the very stones of which the walls of Thebes were built, may, for all we know, have been merely a skilled worker in concrete. Certain it is, however, that long before the dawn of authentic history, at a period back beyond the prying curiosity of scientific research, and while the earth was still young, people lived who knew a great deal about cement and its adaptability to the various needs of construction, as well as its wonderful powers of resistance to the disruptive action of natural forces.

The ancient Egyptians understood the use of hydraulic cement. It has been proved that in some of the marvelous constructions which still endure as monuments of their engineering skill, they used a porous lava possessing hydraulic properties and containing the basic element necessary to the making of cement somewhat similar to the Portland cement of the present day. Many of the sarcophagi in which they placed their dead were made of artificial stone. The majestic pyramids which for over 4,000 years have reared their stupendous forms above the desert, and which still tranquilly laugh defiance at the ravages of time, were built in part of concrete. It is now generally conceded that in the construction of their upper tiers, at least, concrete was the material employed; and the massive blocks of stone that have baffled past ages by the mystery of their transportation to such elevations were probably borne to their destination by the pailful, and formed directly in place. Evidence that these blocks are of man's formation is found in the fact that breakages in some of them have revealed small pieces of wood embedded in the mass.

The Romans constructed many miles of highways and walls and aqueducts from hydraulic cements, which they also used to some extent in the building of their residences and temples.

The word "Concrete" is itself of Latin origin, meaning "grown together" ("con," together; and "crescere," to grow), and implies a body formed by the coalition of separate particles into a solid mass.

Roadbeds of concrete resounded to the thundering tread of the Roman legions as they went out to or returned from their campaigns of world-wide conquest. The famous Appian Way, over which the Apostle Paul entered Rome, as described in the Book of Acts in Holy Writ, was underlaid with cement concrete and topped with paving stones. The latter have been worn away, but the concrete is still intact, just as when the Romans laid it. The aqueducts which supplied the imperial city with water were built without reinforcement, and are still in almost perfect condition. The cement lining of the Pont du Gard at Nismes, in Southern France, a Roman aqueduct built in the first century A. D., is still hard and smooth as when first put in place. The pools of King Solomon, nine miles from Jerualem, were built of concrete—and still furnish water for the city. Many residences of the Roman nobles were constructed of concrete unfaced by brick or stone; and wood framing was used in casting the walls, in much the same manner as wooden forms are used in concrete construction to-day. The Colosseum was built on piers and foundations of concrete; and Professor Middleton, in his work on "Ancient Rome," tells us that the entire upper floor of the Atrium Vesta was formed of one great slab of concrete fourteen inches thick and having a span of twenty feet, supported on edges, but having no intermediate supports. The Pantheon of Rome, a circular temple originally dedicated about the time of Christ, in the reign of Augustus, has a dome 142 feet in diameter, constructed mainly of concrete, and still in perfect condition after 1900 years of service. The Aurelian wall about the city, over ten miles in length, still standing, was built of the same material. An English writer has said: "In strength and durability, no masonry, however, hard the stone or large the blocks, could ever equal these Roman walls of concrete, for each wall was one perfect coherent mass, and could be destroyed only by a laborious process like that of quarying hard stone from its natural bed."

Over other widely scattered portions of the earth, similar evidences are found that the early races were familiar with the constructive use of cement. In Spain and other Old World countries are found abundant examples of the ancient use of concrete in forms that have survived the ravages of time and the elements. The lookout towers of Ireland, supposed to have been built by the Druids more than a thousand years ago, are made of hydraulic cement concrete, cylindrical in form, about six feet in diameter, and 100 feet high. Some years ago one of these towers was undermined and overturned. In its fall, the shock of the huge mass was so great that the shaft sank into the ground to the depth of one-half its diameter for its entire length; yet not the slightest fracture occurred—a fact which demonstrates not only the strength but the elasticity of concrete. Had the shaft been a monolith (that is, one solid piece) of even the best quality of natural granite, the shock of such a fall would have reduced it to fragments.

A similar story as to the early use of concrete might also be written of the vanished races of the New World. The Peruvian builders in the days of the Incas employed concrete, impelled probably by the necessity of adopting a form of construction that would withstand earthquakes and volcanic tremors. Some of their structures, though built many centuries ago, have endured to the present time. And even in our own North America, the use of this most versatile of plastic materials of construction antedates all historic or prehistoric records written in less enduring form. Twenty miles northeast of the city of Mexico are remains of a former but now vanished civilization, in the shape of pyramids of masonry that were built partly of concrete. And ethnologists and antiquarians tell us that as far back as eleven thousand years ago, while the slow-working geological forces had not yet moulded the face of this continent to its present contour, the remarkable race of men known as the Mound Builders, along what are now the edges of the Ohio valley, were accustomed to boil salt water in kettles of artificial stone. And in the manufacture of their pottery, specimens of which are the most enduring mementos of their intelligence, they used as the ingredients of their concrete, clay and carbonate of lime or sand—which are used for similar purposes to the present day.

In one sense, therefore, concrete can lay no claim to novelty, being a return to principles once well known if less perfectly understood than now, but for a long time lost sight of or forgotten. It is not to be supposed, however, that in quality of material or in methods and applianecs of construction the early workers in concrete attained anything like the high standards of technical accuracy that characterize the manufacture of cement and its use in the various forms of concrete construction at the present day. The ancient skill which had been developed in the making and working of concrete appears for many hundreds of years to have become a "lost art." As the empires gradually went to pieces through internal decay, and the world stumbled on toward that dark period of utter standstill known as the Middle Ages, when the lamp of progress flickered but faintly in the cells of monasteries, we find that even in the elaborate cathedrals erected during this period the use of hydraulic mortars had given way to that of fat lime and silt mortars. These ingredients in time became inert powders devoid of strength and hardness, which necessitated constant watchfulness and costly bracing and repairs to keep the structures erect.

About the beginning of the eighteenth century there came a revival in the demand for hydraulic mortars, which was met by supplies of Pozzuolana from Italy, or of the low-grade natural cement known as "trass," from Germany. In the erection of the Eddystone Light, about 1756, John Smeaton developed a form of natural hydraulic cement, after experimenting on limestones containing different proportions of clay, and demonstrated for the first time the practical importance of chemical analysis in cement making.

With the discovery, however, of true Portland cement, the inauguration of accurate physical and chemical tests, and, in particular, the combination of concrete with reinforcing members, and analysis of the principles of constructive design, an entirely new era of development was ushered in—one distinctively modern and without analogy or precedent.

Development of the Portland Cement Industry.—Notwithstanding its great powers of endurance, the concrete employed by the Romans was far inferior to present-day standards. That used in constructing the dome of the Pantheon, for example, was made by mixing broken stone and coarse gravel with a cement consisting of a mixture of slaked lime and volcanic ash. This cement was not a Portland cement, but closely resembles the Pozzuolan cement still produced in small quantities in Italy and other countries of Europe, which is manufactured by mixing volcanic lava (a natural product) and slaked lime, the resulting mass forming a rather weak hydraulic cementing compound without requiring the agency of heating.

Portland cement, on the other hand, is wholly an artificial product, made by calcining or burning a finely ground and carefully proportioned mixture of calcareous (limey) matter with an argillaceous (clayey) or siliceous substance (sand, quartz, or slag) to the point of incipient fusion, and grinding the resulting clinker to an impalpable powder, to which in some cases a small percentage of gypsum or other special ingredient is added.

Although Portland cement now far surpasses all other kinds of cement in structural importance and in extent of use, natural and slag cements still figure to some extent in construction work. The difference in composition, strength, and lasting qualities, however, is great.

The first cements manufactured in modern times were made in England. As early as 1790, one Joseph Parker made a cement by burning lumps of chalky clay stones and finely pulverizing the clinker. This he called "Roman cement," either because it resembled the cement used by the Romans in its property of setting under water, or because of its similarity in color to the lavas found in the vicinity of Rome.

It is, however, to Joseph Aspdin, a brick-mason of Leeds, that the credit of originating the first **Portland Cement** is due, and he is therefore generally recognized as the father of the modern Portland cement industry. After experimenting for some years along the lines indicated by Parker, Aspdin produced a cement by mixing pulverized limestone with clay in certain proportions, calcining in kilns, and grinding very fine. This cement he called "Portland" because of its close resemblance in color to the famous stone quarried on the Isle of Portland in the English Channel, from which St. Paul's Cathedral, the Eddystone Lighthouse of Smeaton, and other prominent structures in England were built.

In 1824 Aspdin obtained a patent on his hydraulic cement. At this time, however, and for years thereafter, nothing was known of the proper proportioning of calcareous to argillaeeous matter, or of the temperature necessary to insure a good product. Many experiments were made by English and French engineers in an endeavor to produce a cement that would command the attention of builders, but no permanent results of value were obtained. It was reserved for a German chemist, in 1828, to formulate the first theory of the action of ingredients and their proper combining proportions to make a true Portland cement. Without this knowledge the securing of reliable material could never be depended upon. Improvements in standard were from time to time effected. Progress, however, was slow; and although the manufacture of Portland cement had been begun in England in 1825, it was not until the middle of the century that this cement secured any wide recognition of its merits as a material of construction. In France, its manufacture was begun in 1846; and in Germany, at Stettin, in 1855; but it was not until 1860 that any considerable quantity was made for the general market in England or on the Continent, and even then but very little was made for exportation.

Portland cement was first brought to the United States in 1865; and during the following decade considerable quantities of the foreign product were imported, constituting the greater portion of such cement used in this country by architects and engineers. In the late seventies, however, the domestic product began to come into competition with the foreign article, and American Portland cement has now almost entirely displaced that of foreign manufacture.

Thus, it was not until some years after the close of the Civil War that the great Portland cement industry in the United States had its beginning. For half a century prior to that time, natural cement—the first kind of cement manufactured in the United States—had practically a monopoly of construction work in this country. Its manufacture had been begun in 1818 by Canvass White, who secured a patent for a Roman cement made from natural rock; and this material was used in the original masonry work on the Erie canal.

The first Portland cement manufactured in the United States was made at Coplay, Pa., in 1872, by David O. Saylor, who exhibited his product at the Centennial Exhibition in Philadelphia, in 1876. This was the small beginning of the present enormous American Portland cement industry. Not, however, until twenty years or more had passed, did the industry in this country begin to show any very substantial increase. Up to 1880, in fact, only six plants for the manufacture of Portland cement had been established in the United States—in Pennsylvania, New York, Maine, Indiana, and Michigan—and the entire American output aggregated a total of only 82,000 barrels. Not only were American methods of manufacture defective as compared with foreign standards, but at that time reinforced concrete was unknown, and the value and various adaptabilities of Portland cement concrete as a structural material were not dreamed of. Natural cement had almost the entire field to itself; and concrete, when it was used at all, was confined almost wholly to foundation and underground work. The slow progress of the Portland cement industry in America during these early days can be seen from a glance at Table I, which gives statistics of production in various years between 1880 and 1895.

It was not until 1896 that the annual domestic production passed the million-barrel mark.

The general commercial application of Portland cement concrete to construction work may

TABLE I

Production of Portland Cement in the United States

Year	Amount Produced	Year	Amount Produced
1880	42,000 barrels	1892	547,400 barrels
1882	85,000 "	1893	590,652 "
1885	150,000 "	1894	798,757 "
1890	335,500 "	1895	990,324 "
1891	454,813 "	....	....

be said to date from 1895. It had previously attained wide use in foundations, and at this time its development was beginning for such strue-tures as dams, piers, breakwaters, sewers, and subways.

Several causes may be assigned for the comparatively slow early growth of the industry, followed by its later marvelous activity. In the first place, the American product, in the early days of the industry, was inferior to the imported German and other European makes. The latter were long looked upon as standard; and it was only after slow and tedious experimentation that the present improved American methods of manufacture were developed so as to produce a high-grade cement which, under the tests necessary for all structural materials, could be depended upon to give uniform and thoroughly reliable results. Moreover, the natural cement industry had attained a flourishing growth, and its competition had to be overcome. Again, the cost of imported Portland cement, on account of the tariff, continued high; and it was only with the gradual improvement in the processes of manufacture that the price asked for a domestic product of a grade equal to if not better than the foreign fell to a figure at which plain concrete could compete with stone masonry, or reinforced concrete with other constructive materials.

As a result, however, of this gradual improvement in quality and lowering in cost, domestic Portland cement has now almost entirely displaced the imported European product in American markets. Striking evidence of this is found in the Portland cement exhibits which were made at the Columbian Exposition in Chicago in 1893, and at the Louisiana Purchase Exposition in St. Louis, Mo., in 1904. At the Chicago World's Fair of 1893, a very noticeable feature was the absence of any American Portland cement exhibit, while Germany had elaborate exhibits of this material. At St. Louis in 1904, these conditions were exactly reversed; and the conspicuous absence of foreign cement exhibits worthy of note served to emphasize the withdrawal of foreign cements from the American market. In 1893, more than four-fifths of all the Portland cement used in the United States was of foreign manufacture. Within the following decade the importations fell off about three-fourths. From one of the smallest, the United States has taken foremost rank among the largest Portland cement producing countries in the world. And records compiled from tests made by government and private engineers warrant the claim that the leading American brands of

Portland cement are superior in quality to any of the foreign cements manufactured at the present day, this superiority being due to unsurpassed raw materials and the high grade of technical knowledge and skill embodied in American machinery and processes of manufacture.

In addition to improvement in quality and lowering of cost, a most important factor that contributed to the rapid expansion of the Portland cement industry in the United States was the application of concrete to steel structural work in fireproof building, and the subsequent development and widespread adoption of what is known as the **Reinforced Concrete** type of construction. These quickly created an unprecedented demand for good Portland cement.

The general use of concrete on steel structures, about 1895, opened up a wide field of new opportunities for the use of cement. This field was enormously expanded by the general acceptance, about 1900, of reinforced concrete as a type of construction of practically unlimited possibilities. The subsequent history of the Portland cement industry in the United States has, in fact, coincided with that of reinforced concrete, one being but a reflex of the other in its unexampled growth.

The invention of the twisted bar for reinforcement was made in 1885 by Ernest L. Ransome, of San Francisco, and about 1890 came the introduction of expanded metal and the mesh systems of concrete construction, which allowed

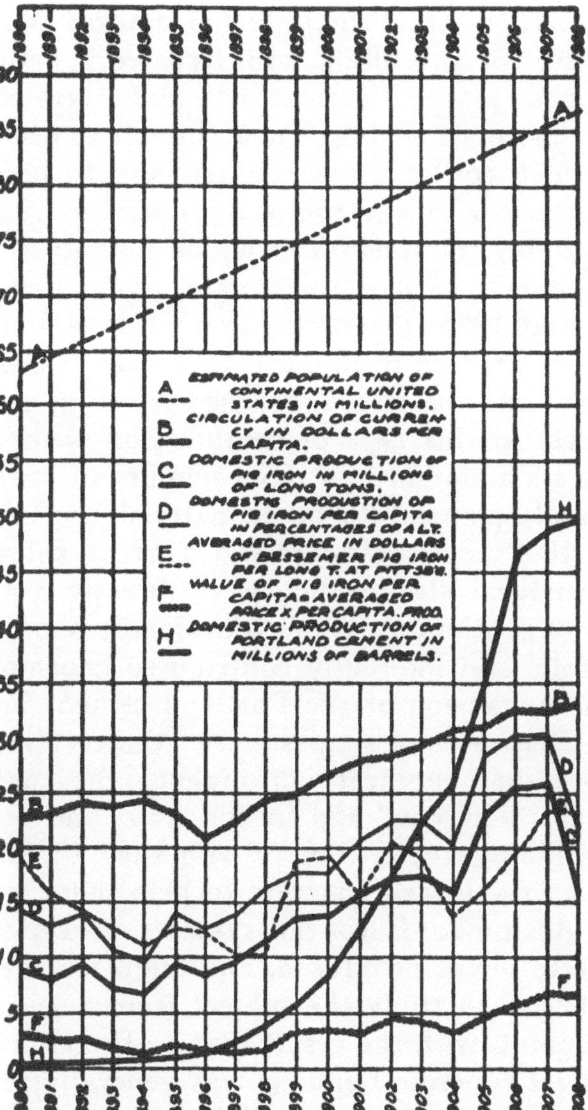

Fig. 1. Diagram Showing Remarkable Advance of Portland Cement Manufacturing Industry as Compared with Other Industrial Features of the United States.

architects to retain their steel structures and use concrete for the filling and arch work between beams, in place of brickwork and terra-cotta. These methods enabled the architects to use concrete in a conservative manner, and to become acquainted with its structural properties, without abandoning the steel frame which constituted the main structural basis of their design. Between 1890 and 1895, the production of Portland cement progressed from 335,000 barrels to 990,000 barrels; and the lessons learned from the use of concrete in this type of combination with steel led to its adoption in a large number of projects originally planned for other material.

In 1897 and 1898 occurred a great shortage in structural steel, which gave directly a great additional stimulus to the adoption of reinforced concrete, and indirectly contributed enormously to the expansion of the Portland cement industry. Deliveries of shapes were so uncertain and remote that engineers throughout the country were at their wits' end to find ways and means for fulfilling their designs, and they turned at once to reinforced concrete to help them out of their dilemma. European systems of reinforced concrete were introduced, and engineers generally began to study the subject from a scientific standpoint, with the result that by the year 1900 the steel concrete building was generally recognized as a structural possibility, and the adoption of reinforced concrete was considered for every conceivable kind of problem.

The statistics of Portland cement production during the years that immediately preceded and followed the opening of the new century, demonstrate in an astonishing way the increasing popularity of concrete and the confidence of the public in its properties. Between 1896 (when the production of American Portland cement first passed the million-barrel mark) and 1908, the country's annual output increased fifty-fold. The 26 plants operating in 1896 grew in number during the following decade to over 100, representing an invested capital of more than $100,000,000. The phenomenal development of the industry will perhaps be best apparent from a glance at Table II, giving statistics of production from 1896 to 1909.

A graphic illustration of the remarkable growth of the Portland Cement Industry in the United States, as compared with the advances made in other industrial lines, is given in the accompanying diagram (Fig. 1), which is based on a chart prepared by Mr. John Birkinbine, of the Birkinbine Engineering Offices, Philadelphia. The curves also afford comparison with the increase in population, monetary circulation, etc.

Beginning at the bottom of the chart with less than a million barrels in 1890, the cement line is gradual in rise until 1896, when its upward tendency becomes pronounced, advancing with such rapidity that it soon crosses that of pig iron, extending far beyond all comparative lines, and finally ending in the great total of sixty million barrels in 1909. It is entirely unique in that it does not show the least downward tendency at any point, a record no other industry has made.

Analysis of the figures of production shows that the output of Portland cement in the United States has multiplied itself nearly ten times in as many years—a truly marvelous growth. The greatest single gain was in 1906, when the production increased over 11,000,000 barrels, and its valuation $17,000,000, over the year preceding. Within a period of five years ending in 1908, the production was more than doubled, the increase in valuation being proportionate. The output during the last four years alone of this period exceeded by nearly 75,000,000 barrels the entire production credited to the thirty years preceding; and the valuation of the product for the same period was over $70,000,000 in excess of the value of the entire production from 1870 to 1904. An examination of the statistics also shows a gradual downward trend in the prices of the finished product.

The prodigious growth of Portland cement manufacturing, especially during the past six years, has been unequaled elsewhere in the annals of American industrial development. But even these colossal figures show but little of the depth or extent of public interest that has been aroused in the subject of cement and concrete construction. Even in periods of business depression, the demand for cement in the building

TABLE II

Portland Cement Production in the United States

Year	Amount Produced	Valuation
1896	1,543,023 barrels	
1897	2,677,775 "	
1898	3,692,284 "	
1899	5,652,266 "	$ 8,074,361
1900	8,482,020 "	9,280,525
1901	12,711,225 "	12,532,360
1902	17,230,644 "	20,864,078
1903	22,342,973 "	27,713,319
1904	26,505,881 "	23,355,119
1905	35,246,812 "	33,245,867
1906	46,463,424 "	61,240,652
1907	48,785,390 "	53,992,5.1
1908	51,072,612 "	55,000,000
1909	60,000,000 " (approx.)	60,000,000 (approx.)

trades has been such as to tax the capacities of the mills to their utmost. Portland cement has long since passed the stage where it was regarded as suitable only for the building of underground and under-water work. It has entered aggressively the field of general construction; and in spite of the fact that not all its structural applications and systems have yet been standardized, nor all its problems solved, it has abundantly "made good" in setting at rest all doubts as to its marvelous adaptabilities and fitness for constructive work in general. In the business and manufacturing districts and the residence streets of cities and towns, along the water-fronts of lake and ocean ports, along the railroad lines and country highways, at summer resorts, in rural villages and on the farm, structures of every size, shape, and style of adornment may now be seen in constantly increasing numbers—from the skyscraper to the one-story garage, from palatial residence to simple cottage or bungalow, from arched bridge to culvert and sidewalk and curb, from the great manufacturing plant to the simple shop, from barn to dog-kennel and hitching post—in whose construction cement has played a leading part. And when the constructive possibilities of this versatile material come to be fully realized by the vast army of the agricultural and so-called "middle" classes—the substantial home-seekers and home-makers who constitute the backbone of our country—the output of the cement mills must show a still greater increase, which will dwarf even the present handsome figures of production.

Dictionary of Terms

Like every other branch of human activity, the concrete industry has developed a list of terms of special meaning. With the usage of these terms, the more common of which are given below, the student of cement and concrete should at the very outset make himself thoroughly familiar.

Aggregates—The solid and relatively coarse ingredients which are bound together by cement in a mass of concrete. In the great majority of cases, the materials used as aggregates are mineral, comprising sand, gravel, crushed stone, furnace slag, cinders, etc., of varying degrees of fineness. In the form of concrete known as "pulp concrete," sawdust, an organic material, is used as an aggregate.

Chemically the aggregates are absolutely inert or inactive in the formation of concrete, playing only a physical part because of their presence as space fillers. Sand or other material fine enough to pass through a sieve of one-fourth-inch mesh, was formerly not classed as an "aggregate," this term being confined to the coarser grades of material. Inasmuch, however, as the function of the sand is essentially the same, both physically and chemically, as that of the coarser ingredients, the later approved practice has come to include sand also among the "aggregates." The word is of Latin origin—from "ad" (to) and "grex" (a flock)—and means things collected together in a mass.

Argillaceous—Consisting of, or containing, clay.

Calcareous—Consisting of, or containing, calcium carbonate or carbonate of lime.

Calcining—Burning; subjecting to sufficient heat to cause partial chemical disintegration, or incipient or total fusion or vitrification.

Cement—The binding material which holds the aggregates together in concrete in a solid mass. Derived from Latin "caedere" (to cut), and signifies any substance used for making bodies adhere to each other, or binding together those which have been cut or broken apart, as mortar, glue, etc. The term was originally applied to rough, unhewn blocks of stone used in construction; later, to marble chips used in making mortar; later, to mortar itself; and was finally extended to include any sort of plastic bonding material.

Center—A middle point, line, or plane, used, for example, in measuring the distance from one joist, beam, or column to another; the temporary wooden or metal supports used in the erection of arches or other superimposed or overlying constructive work.

Constancy of Volume—See Soundness.

Core—The mould used to form the hollowed-out part of a cement or concrete block.

Crazing—The checking or cracking of the surface of artificial stone, concrete, etc.

Efflorescence—The white or grayish crust sometimes formed on the surface of brick, concrete, stone, etc., due to the leeching-out of soluble chemical salts.

Empirical—Based on practical experience as distinguished from refined scientific and theoretical calculations. Empirical rules, for example, are founded merely on the observed results of actual experience, while the theory underlying such rules may never yet have been formulated.

Erosion—Wearing away, as, for example, through the action of water, of the weather, etc.

Failure—Collapse, rupture, or fracture—as the failure of a beam or column.

Forms—Wooden or metal constructions set up temporarily for the purpose of holding concrete in shape and in place until it has set and has hardened sufficiently to be self-supporting.

Frost Line—The average level to which frost penetrates into the ground. It varies in different localities.

Fusion—Melting; usually applied only to melting of mineral substances which takes place at relatively high temperatures.

Gauging—Stirring or mixing.

Granolithic—Consisting of artificial stone of fine granular structure.

Grout—Thin mortar.

Gypsum—A mineral consisting of the hydrous sulphate of calcium—that is, sulphate of lime containing a considerable amount of water of crystallization. When the water is driven off by heat, the resulting product is "plaster of Paris."

Hardening—See Setting.

Hydraulic—Active in the presence of, or under the influence of, water. Hydraulic lime and hydraulic cement, for example, are lime and cement that will set and harden under water. Derived from Greek "hudor" (water).

Incipient—Beginning to take place. Burning to "incipient" fusion, for example, means heating to a temperature at which fusion or melting of the materials begins.

Monolith—A term applied to single-piece work, or walls built with concrete between frames. A word of Greek origin; its literal meaning is "one stone."

Mortar—A plastic mixture of lime, cement, plaster of Paris, or other material, with sand, water, hair, etc.; used for plastering or for binding together stone, brick, etc., in construction work.

Plaster of Paris—Anhydrous sulphate of lime. Produced by calcining gypsum at a temperature sufficient to drive off the water of crystallization. Mixed with water, it forms a paste which quickly hardens, and is much used for casts, mouldings, hard plaster, stucco-work, etc. So called because first brought from a suburb of Paris.

Plastic—Capable of being moulded, formed, modeled, or spread, like mortar or paste.

Retemper—To stir thoroughly again, so as to give a workable consistency; to reguage, as in stirring up mortar or concrete that has already begun to set.

Rubble—Rough, broken stone used in coarse masonry, or to fill between the facing courses of walls, or in making up the interior of massive stone or concrete work.

Screeds—Levels for guides and bearings for leveling and ruling off.

Setting—Loss of plasticity; solidification from plastic state, due to crystallization of the cementing material. Takes place soon after mixing. To be distinguished from "hardening," which may continue for months or years after the "set" is complete.

Siliceous—Containing silica (quartz or sand), or partaking of its nature. Written also "Silicions."

Soundness—Absence of tendency to cracking, swelling, shrinking, distortion, or disintegration under varying conditions of moisture and temperature; constancy of volume.

Striking—Taking down and removing—as, for example, striking the centers or forms used in arch construction.

Stucco—Cement mortar or gypsum plaster, used in exterior finishing of buildings.

Subfoundation—The first layer of material placed in excavated ground for sidewalks, floors, etc.; usually it is cinders.

Tamping—The pounding of concrete to reduce voids and force the aggregates as closely together as possible into a compact mass.

Tensile Strength—The holding power or measure of adhesiveness of concrete or cement mortar; power to resist the action of forces tending to pull apart. Contrasted with **Compressive Strength**, the power to resist crushing under direct pressure.

Trap Rock—A dark-colored igneous rock of great weight and strength, including basalt, feldspar, etc.

Vitrify—To make glass-like, as vitrified clay, glazed surfaces, etc.; usually done by heating to fusion.

Voids—The air-spaces or interstices between the particles of aggregates.

Materials and Manufacture of Concrete

What Concrete Is.—Concrete is an artificial or manufactured stone consisting of aggregates (sand, gravel, etc.) and cementing materials, mixed together with water. The ingredients to be used, as well as their proportions, vary with the kind of work for which the concrete is intended.

If a mixture could be made containing absolutely no voids or air-spaces—a mixture in which every part of the surface of every particle was covered by cement paste in intimate contact with it, and every intervening space between the particles completely filled with the cement paste —that mixture would be an **ideal mixture.** And provided the ingredients in this ideal mixture were of the kind and size and proportions best adapted to the work in view, and were of uniform good quality throughout, the resulting mass would be an **ideal concrete.**

A concrete which would be ideal for one purpose or location might not be at all suitable for another. The particular use that is to be made of the concrete is a factor which must always be considered in the selection and proportioning of its ingredients.

In practice it is impossible to attain the absolute perfection of these theoretical ideals. The air-spaces in concrete can be reduced in quantity, but cannot be entirely eliminated. Even with the utmost practicable care and thoroughness in proportioning and mixing, and notwithstanding the use of enormous power in compressing the mass, some small, infinitesimal voids will still remain. The fallibility, also, of human perception and judgment in the choosing and testing of materials, and the greater or less imperfection that characterizes all devices and processes of man's making, will introduce at least some variation from the theoretical standards of absolute perfection. An ideal concrete is no more possible in actual practice than is an ideally perfect wood or stone or brick or steel or other structural material. We may know perfectly well what the ideal requirements are, but their complete realization will always be beyond us. The utmost that we can do is to approximate them more or less closely. They are the goal to be aimed at in all endeavor, and our work will be perfect only in proportion as we approach them. Nature herself affords us the only examples that are to be found of absolute perfection—in the handicraft of the Master Workman.

In spite, however, of theoretical limitations, and the fact that not all the problems of concrete have yet been finally worked out, nor all its details standardized, it can be said that the modern art of working in concrete has now reached such a stage of development that a degree of perfection sufficient for all practical purposes is quite within the reach of everyone who will give the matter a little careful study.

The wonderful adaptability of concrete to various forms and combinations of construction is based on its property of changing from a plastic condition—in which it can be moulded to any shape or design—into a firm, rigid, rock-like mass, through the setting and hardening of the cement paste. No other building material that the world has ever known can compare with concrete in this respect. In addition, however, to this peculiar property, concrete possesses certain qualities which have irresistibly compelled the structural world to study carefully its alleged advantages as a substitute for wood, cut stone, and other old-time building materials, and which have naturally caused its claims to be looked upon with more or less favor. These qualities are its durability and accumulative strength, its non-corrosion in moist places and freedom from rotting under any conditions, its cleanliness and sanitary value, its resistance to fire and the action of the elements, its protection of steel from corrosion, its deadening to sound, its ease of manipulation, and its moderate and steadily lowered cost as compared with the increasingly higher cost of lumber, due to the gradual exhaustion of our natural resources of forest wealth.

KINDS OF CEMENT

Most of the cement now employed in constructive work falls under one or the other of two main heads—**Natural Cement** and **Portland Cement**, the latter being much the more extensively used, especially where great strength and absolute reliability as to uniform quality are required. In some instances, the so-called **Pozzuolana** and **Artificial Slag Cements** are also used; and, for certain special locations and purposes, a special kind of cement known as **Keene's Cement** is preferred.

NATURAL CEMENT

Natural cement is so called because the raw material from which it is made is found in a **natural** state and is used just as it comes from the earth, requiring neither artificial proportioning nor the admixture of other ingredients.

In the case of the Portlands, on the other hand, the raw material is in every case an **artificial** mixture. Some natural ingredients, it is true, are used in the making of Portland cement; but the proportioning and mixing are always artificially done, and only after careful chemical and physical tests to determine exactly the proportions and the degree of fineness of grinding necessary for developing the qualities desired.

Natural cements are made by calcining **cement rock**—argillaceous (clayey) or magnesian limestones (dolomites). The temperature required is considerably lower than that needed for the making of Portland cement. The resulting clinker is ground to a fine powder, which is then immediately ready for use.

Natural cements vary in color from light to dark gray, according to the character of the stone from which they are made. They are marketed under various names, such as "Rosendale," "Utica," "Louisville," "Akron," "Milwaukee," "Fort Scott," etc., depending generally on the locality where the cement rock is found. The cement, for example, known as **Rosendale**—a name sometimes applied loosely to all natural cements—is made by burning magnesian limestone, of which very important deposits occur near the town of Rosendale, in Ulster County, New York. After being quarried and sorted, the stone is heated in large kilns. The over-burned and under-burned portions are carefully discarded, and the properly burned portion is passed at once to crushers, where it is reduced to a fine powder, then being carefully screened and packed for shipment in either barrels or bags.

Owing to the fact that the chemical contents of the cement rock may differ greatly even in different parts of the same quarry, the strength of the product may show great variations on test; and it is only by careful selection of the desirable strata that a uniformity of quality can be insured. Much of the disfavor with which natural cements have been looked upon by engineers has been due to this uncertainty and variation in strength. In the case of Portland cement, on the contrary, a substantially absolute uniformity of composition and consequent reliability as to strength in any particular brand, are not only attainable, but are practically assured by the careful technical supervision exercised over all the processes and details of manufacture by the leading American Portland cement makers.

In strength, natural cement falls considerably below Portland. Its hydraulic activity, on the other hand, is considerably greater; that is, it sets much more quickly than Portland cement—which may or may not be an advantage, depending on the time available and whether a quick or slow setting is desirable. The setting of natural cement can be retarded, and thus regulated to some extent, by the addition of a small proportion of lime in mixing the mortar or concrete.

For use under sea water, a good quality of natural cement seems better adapted than the ordinary Portland. The former is of constant volume in sea water, probably due to the absence of free lime and to the presence of a larger amount of magnesia than can safely be used in Portland cement; while the latter has a tendency to expand or swell and disintegrate, probably because of the chemical action of the sulphates in the water upon the constituents of Portland cement, and the consequent formation of soluble salts.

Natural cement costs less than Portland, and for this reason is still used to a large extent in construction where no very great degree of tensile strength is required. It is well adapted, for example, for use in the interior of heavy masoury. The lower cost of natural cement as compared with Portland is explained by the fact that its raw material is wholly a natural product used just as it comes from the quarry, and in its manufacture a much lower temperature is required for burning than is needed for the making of Portland cement.

Natural cements will not carry so large a proportion of sand with good results as will Portland, and they require more water than the latter for proper hydration; they are more plastic, however, and work more smoothly under the trowel or other tools.

The Western natural cements, such as "Louisville" and "Akron," weigh 265 lbs. to the barrel; but "Rosendale" from New York or Pennsylvania weighs 300 lbs. to the barrel. Natural cements are thus lighter than Portlands, which weigh about 380 lbs. per barrel; and they are usually sold in cloth bags, three bags to the barrel, instead of four bags, as is the case with Portland cements.

For many years, as already noted, natural cement enjoyed a practical monopoly of the constructive field in concrete work. In recent years, however, the natural cement industry has shown a remarkable decline. This is readily explained by improvements in American machinery and by skilled technical supervision of the processes of manufacture which have resulted in the production of absolutely reliable Portlands; by the greater adaptability of the latter to constructions requiring the utmost possible strength; and by the gradual lowering that has been effected in the cost of Portland cement.

SLAG CEMENT OR POZZUOLANA

The use of slag cement dates back to ancient times. The cement used by the Romans was made for the most part from a simple mixture of lime and volcanic lava. The lava first used for this purpose appears to have been obtained at Pozzuoli, a village lying near the base of the volcano Vesuvius. For this reason the name **Pozzuolana**—or, as it is very commonly spelled, **Puzzolan** cement—was first given to this form of cementing material, and has since been extended so as to include all forms of slag cement which are made without calcining.

In Italy and some other parts of Europe, Pozzuolana is still made from lava; but in America, **furnace slag** is used instead. This material is simply mixed mechanically and thoroughly with slaked lime. No burning is required, but the mixture is ground exceedingly fine.

In chemical composition, slag cement consists of a combination of silica and alumina, mixed with hydrated lime. Like Portland cement, natural cement, and hydraulic lime, it has the property of hardening under water; but in hardness and strength it falls far below the standard of true Portland cement. It is of a pinkish hue in color, and of lower specific gravity than Portland cement. It is not used to any great extent in building operations. It cannot be depended on in freezing weather, and is not well adapted for dry work which is to be left exposed above ground to the action of the atmosphere. Its principal use is for underground work and in moist locations. Sea water appears to affect it less than it does Portland cement; and for that reason slag cement is sometimes used instead of Portland in cases where large masses of masonry or concrete are to be deposited under sea water, and where great strength is not the prime consideration.

PORTLAND CEMENT

In 1908, Portland cement was manufactured in at least twenty-five of the forty-six States of the Union. The raw materials used by the different plants vary considerably, so that no absolutely universal rule can be laid down for the selection or proportioning of the ingredients from which Portland cement is made. It is equally impossible to give any exact formula for its chemical composition, since the latter, too, is found to vary through a considerable range. Nor, in fact, is it even possible to give a definition of Portland cement which will invariably afford an accurate guide as to its physical ingredients or chemical composition. These things are not secret mysteries; but they are variable factors, no two brands being exactly the same in these respects. Thus, Portland cement eludes all attempts to reduce it to the uniformity of a positive chemical formula or universal standard of composition.

It is not to be supposed for a moment, however, that a true Portland cement is any uncertain quantity or unworthy of implicit confidence as a structural factor. Its composition may vary, and does vary; but that is immaterial. From the structural point of view, the important thing is that the cement, and the concrete made from it, shall meet certain requirements as to strength, durability, constancy of volume, fire resistance, etc.; and in these respects, not only can Portland cement be manufactured with such perfection as to conform to all the positive standards that experience has shown to be desirable; but in construction work—especially where failure would involve great destruction to property and possibly to human life—it should be the invariable practice to make assurance doubly sure by careful testing, to see that the particular brand of cement whose use is proposed does actually come up to standard requirements in every respect.

What Portland Cement Is.—So far, then, as it is possible to give an accurate definition, Portland cement may be defined as any cement which, on being tested, will manifest the characteristic properties that engineers have agreed to associate with the name "Portland," and which will meet the requirements that have been accepted as standard. If, for example, from even such unpromising material as eggshells and street sweepings, a product could be obtained which would pass all the standard tests, it would be fully entitled to be called a "Portland cement."

Raw Materials.—As a matter of fact, however, the ingredients from which true Portland cements can be manufactured, although very widely distributed in nature, appear to include only a comparatively limited list. In the main, they comprise calcareous or limey materials, argillaceous and siliceous materials, and blast-furnace slag, with minor accessory materials added for the purpose of regulating the activity of the product.

The forms of calcareous matter generally used are three in number—**limestone, chalk,** and **marl**—each of which must contain the requisite amount of lime; for about sixty per cent, or nearly two-thirds by weight, of all Portland cement consists of this ingredient. Limestones are found almost everywhere. Chalk and marl are forms of limestone, usually of a purer calcium content than the ordinary "limestone" so called. Marl is a finely powdered substance usually found at the bottom of lakes or in swamps, having been deposited there by the action of water; and consists almost entirely of pure carbonate of lime. It is dredged out of its bed or mixed with water, and pumped into tanks for future use, being employed in what is known as the "wet process" of cement making.

Clay, shale, and **slate** are used for the silica, alumina, and iron contents of the cement; and **blast-furnace slag** is also used for this purpose.

In physical form, Portland cement is an exceedingly fine powder, ordinarily of a dull bluish or greenish gray tint, though brands of pure "white" Portland are now also made.

Manufacturing Processes. Portland cement is obtained by first grinding and mixing together, in carefully determined proportions, the principal raw ingredients, then heating the mixture to a temperature at which melting begins (2,000° to 4,000° F.), and finally grinding the resulting **cement clinker** to an impalpable powder, a small percentage (not over three per cent) of gypsum or other material being subsequently added in some instances, to retard the setting or otherwise regulate the activity of the product.

All the Portland cement mills have chemists in charge of laboratories, who determine the proportions of the materials that are being used from day to day, and thereby maintain a standard of mixture as nearly perfect as possible.

The degree of fineness to which the clinker is ground has an important bearing on the strength of the cement. The finer the cement, the higher the test it will stand, both in tension and in compression. The extreme degree of fineness considered necessary for the best results may be seen from the fact that in the case of the leading standard American brands of Portland cement, the manufacturers guarantee that 95 per cent of their product will pass through a sieve having 10,000 meshes to the square inch, and 80 per cent through a sieve having 40,000 meshes to the square inch.

In the mixing of the materials for Portland cement, two processes are used—the **wet** and the **dry**—depending on the amount of water present. Where marl is the calcareous ingredient, the "wet" mix is almost invariably employed. In either case, the materials are first crushed and mixed, and then finely ground, whereupon they are ready to be passed to the calcining kilns.

The Rotary Kiln. In the early days of the Portland cement industry, a simple, vertical kiln like those used for lime burning was used. This type of kiln, however, while fairly efficient as to fuel cost, required too much labor, and the output was small. The result was the development of the **rotary** or **revolving kiln,** which is the type of kiln almost universally used at the present day.

The rotary kiln was evolved only after much experiment. As at first perfected, it was applicable only to "dry" mixtures, and fuel oil and gas were the only fuels that could be used. Rotary kilns are now, however, used for calcining both "wet" and "dry" mixtures; and in many cases powdered coal, blown in by forced air-blast, is employed as fuel. In the making of "white" Portlands, an oil or gas flame is still used, in order to avoid staining the product.

The rotary kiln consists of a steel cylindrical tube, five to eight feet in diameter. It is lined, except at its upper end, with a highly resistant firebrick, to withstand both the high temperatures generated and the destructive action of the molten clinker. Its length, for "dry" materials, is 60 to 150 feet; and for "wet," 80 feet or longer. In its mounting, the tube is set with one end higher than the other, so as to give it a gentle incline—about one-half inch to the foot. It is mounted on geared bearings which revolve it slowly on its axis. The fire is applied at the lower end. The mixture of the raw materials is fed at the upper end; is heated to the melting point as it passes slowly down the tube; and is discharged as cement clinker at the lower end, being then ready to pass to the final grinding.

A description of the working of a Portland cement plant—which, with minor differences of detail, may be considered as a typical example—is given by one of the leading American manufacturers as follows:

"Large bins approached by elevated tracks at one end of the raw material mill, provide the storage room for the raw material. From these bins, the limestone is fed directly into gyratory crushers, which prepare it for the driers, into which the stone then enters. The slag is also fed into separate driers, all of which are maintained at a sufficiently high temperature to drive off all the moisture. After passing through the driers, each of the raw materials is conveyed into separate revolving ball mills, where the rolling and impact of forged steel balls reduces them to extreme fineness. The two materials are then carried to separate hoppers, from which they are drawn off and properly proportioned by electrically operated weighing devices. The mixture of slag and limestone, already a powder, is then passed into the tube mills where flint pebbles rolling in a steel drum thoroughly mix it and complete the process of pulverization.

"The next step is the calcination of this raw mixture in rotary kilns. Leaving the rotary kilns, the clinker produced by the burning is dropped into storage pits, where it is cooled and cured. Electrically operated cranes then carry the clinker to the finishing mill, where it is passed through jaw crushers to Kent mills, and then over Newaygo separators. After the addition of a certain quantity of gypsum, which is done by means of an automatic device, the material again passes through the tube mills similar to those used in finishing the raw material. Being now ground to a specified degree of fineness, the cement passes on to a belt conveyor, and is distributed into the various bins of the storage house.

"The cement, in passing from the finishing mill into the storage house, is automatically sampled, from which samples the various tests for quality are made in the laboratory."

From the above descriptions of the raw materials and the processes of manufacture, the reader will be able readily to grasp the full significance of the following "official" definition of Portland cement:

As defined by the American Society for Testing Materials, Portland cement is:

"The finely pulverized product resulting from the calcination to incipient fusion, of an intimate mixture of properly proportioned argillaceous and calcareous materials, and to which no addition greater than 3 per cent has been made subsequent to calcination."

It will be seen that no absolute standard for the chemical composition of Portland cement can be formulated. A good commercial Portland cement, however, will conform approximately to the following chemical analysis:

Lime ($Ca\ O$)—57 to 65 per cent.
Silica ($Si\ O_2$)—21 to 24 per cent.
Alumina ($Al_2\ O_3$)—6 to 8 per cent.
Sulphuric acid, anhydrous (SO_3)—0.5 to 2 per cent.
Oxide of Iron ($Fe_2\ O_3$)—2 to 4 per cent.
Magnesia ($Mg\ O$)—1 to 4 per cent.
Water and carbonic acid—1 to 3 per cent.
Sulphur, as sulphides—None.

Portland cement is packed for shipment in paper bags, cloth sacks, or wooden barrels. The cloth sack gives the most conveniently handled package for the average user. It costs a little more than the others, but this is offset by a rebate allowed for the return of the empty sacks. The bags must be kept dry and untorn, and returned by freight exactly as directed by the cement company. Paper bags are easily torn, and cause a big percentage of loss especially on small jobs where any carrying has to be done. Barrels are too bulky to be handled easily, and are too large a unit for measuring.

The weight of the shipping units of cement varies slightly; but in general, a paper or cloth bag contains 95 lbs. of cement, and four such bags make a barrel of 380 lbs.

Storage of Cement. Since cement readily absorbs moisture from the atmosphere, becoming lumpy or even a solid mass—in which condition it is worthless and must be discarded—**it must be stored in a dry place.** If an attempt is made to break up lumpy cement and use it, it will be found to have lost most of its adhesive power and hardening qualities, and so to have lost the greater part of its strength and value as a building material.

In the store-room, the cement should not be piled directly on the floor. Blocks should be laid on the floor, and planks placed over them, thus providing a platform slightly raised above the store-room floor. On this platform, the cement should be piled, and a covering of canvas or roofing paper should be thrown over the pile.

If properly stored, where there is no absorption of moisture sufficient to harden it, Portland cement can be carried from one season to another without deterioration. Some makes, in fact, have shown improvement. Long storage of cement tends to retard its setting, though not preventing its ultimate acquisition of full hardness and strength.

Setting and Hardening. When cement is ready to be used, and is mixed with a sufficient amount of water to bring it to a paste, it soon loses its plastic nature, and finally reaches a point at which it can no longer be distributed without producing a rupture. This change of condition is known as **setting,** and should be distinguished from the **hardening** of the mixture. Setting is usually complete within a few hours, whereas hardening may continue for months or even years. A further distinction is made between **initial setting** and **final setting.** The time required for setting will depend largely on atmospheric and weather conditions, as setting proceeds more slowly in cold weather than in warm. The initial set generally takes place within thirty minutes, and final setting within four to five hours.

Keene's Cement

Keene's cement is a special form of gypsum plaster which has certain unique and valuable properties. It is much harder than ordinary gypsum plaster, will take a good polish, and can be washed with damp cloths without injury. For these reasons it is often used for wainscoting, column coverings, casts, mouldings, etc., intended for places of unusual exposure, common gypsum plaster being hard enough for all ordinary conditions. Hydrated lime is often present in the American product, but imported Keene's cement is generally pure.

This form of cement is made by saturating lump gypsum which has been subjected to intense heat, in a solution of one part alum to thirteen parts water. The rock so treated is ground, then treated again wth the same alum solution, and finally ground again.

CONCRETE AGGREGATES

While cement, in concrete construction, is a very important element, nevertheless the other materials with which it is combined, and the manner of proportioning, mixing, and depositing them, are also factors of prime importance, and should be submitted to the same carefulness of inspection, preliminary testing, and intelligent supervision of competent authority as may be required of the cement itself.

The solid materials which are mixed with cement and water to form mortar and concrete, are known under the common name of **Aggregates.** The substances chiefly used as aggregates are sand, crushed stone, and gravel. The stone may range from hard trap rock and granites to the weaker and more porous varieties of sandstone, etc., and the list of available materials includes also shells, slate, shale, cinders, slag, crushed lava, broken brick, metal filings, and—for "pulp" concrete—sawdust.

A fairly good concrete can be made with almost any mineral aggregate, **provided enough cement is used.** In order, however, to save cement, reduce cost, and at the same time secure the best results, it is exceedingly important that the concrete worker know how to select the best aggregates for the work in hand, and how to determine the proper proportions to use of fine and coarse ingredients.

Selection of Aggregates

In the first place, an aggregate for any kind of work should be **clean.** Dirt is a sign of weakness. Any foreign substance, such as mud or clay, forming a coating on the particles of the aggregate, prevents the cement from coming into contact with the surface, and lowers the cohesive strength of the concrete, or its power to hold itself together. It may also seriously retard the setting. A loss of 15 or 20 per cent in strength may be caused by the use of a dirty aggregate.

Coarse aggregates can be cleaned by spreading them on an inclined plane, turning a hose on the pile, and allowing the water to carry away the impurities. Fine aggregate can also be cleaned in a similar way, by spreading on an inclined platform in a layer 3 or 4 inches deep, and washing gently with a ¾-inch hose from the high end of the platform. Pieces of board projecting up about 4 inches above the face of the platform, should be nailed around its edges to confine the sand, the water being allowed to flow over the top of the board nailed across the low end. A small quantity of clay or loam will not injure the sand, but any amount over 5 per cent should be washed out.

It may sometimes be advisable to use the aggregate that is nearest at hand, and to depend for results on an extra amount of cement. A natural bank of sand and gravel, for example, may be located near the work, giving an abundant supply of aggregate at a cost of very little more than the labor required to get it out. These natural mixtures, however, ordinarily contain an excess of sand, and best results are usually secured by screening, and remixing in the proper proportions.

In the selection of aggregates, it is well to bear in mind that the object is to make a mixture in which all the voids or spaces between the particles shall be filled. The voids in the stone or gravel should be filled by the sand, and the voids in the sand should be filled by the cement paste. This result is secured by varying the proportions of coarse and fine material. Since a perfect filling of the voids is rarely if ever attained, it is well to use a little more sand and a little more cement than would be just enough to fill them.

The volume of concrete will not be equal to the combined volumes of the cement and aggregates, but will be considerably less, because the finer particles hide themselves, as it were, in the spaces within the mass of the coarser particles. In this way, the pile of concrete resulting from a mixture of cement, sand, and gravel will be only slightly larger than the original pile of gravel. Six barrels of gravel, three of sand, and one of cement, for example, will not give ten (1+3+6) barrels of concrete, but will give only a little more than six barrels.

The volume or yield of the concrete affords, in fact, one method of determining the best proportions of aggregates. If the proportion of cement is kept the same, that mixture of coarse and fine aggregates will be the best which will give the smallest volume of concrete from a certain weight of aggregate. This test is called the **yield** or **volumetric test**, and gives a rough but very quick and convenient indication of the relative density and strength of the concrete resulting from different mixtures of aggregates. The smaller the volume, the greater the density and strength, and the better the quality of the concrete.

There is the closest relation between the density and the strength of concrete. Even a slight increase in weight per cubic foot will add very decidedly to the strength. In this connection the importance of the coarser aggregates cannot be too strongly insisted upon. The use of coarse material is essential to density, since coarse materal contains the smallest amount of voids. Different kinds of sand, gravel, and stone vary greatly in the extent of their voids; and by judiciously mixing coarse and fine materials, the voids may be much reduced, and the weight and density of the concrete increased.

If sand be screened so as to take out the coarse grains, the voids will be increased and the weight reduced, thus injuring the sand for making concrete. Strength may be improved by adding coarse material, even though the proportion of cement is thereby reduced. This has been repeatedly shown by experiment. A mixture of cement and sand alone will form a rather weak concrete, especially if the sand is fine. By adding gravel—say about twice the quantity of gravel that there is of sand, or a little more—a concrete will be obtained containing, of course, a greatly reduced percentage of cement, but of greatly increased strength.

In selecting aggregates for concrete, then, it may be laid down as a general rule that those should be chosen which give the greatest density. If it is practicable to mix two materials, as sand and gravel, the proportion which gives the greatest density should be determined by experiment, and rigidly adhered to, whatever may be the proportion of cement it is decided to use. Well-proportioned dry sand and gravel, or sand and broken stone, well shaken down, should weigh at least 125 lbs. per cubic foot. Limestone screenings, owing to minute pores in the stone itself, are somewhat lighter, though they may give equally strong concrete. They should weigh at least 120 lbs. per cubic foot. If the weight is less, there is probably too much fine dust in the mixture.

Selection of Sand. For a rich mortar—that is, one relatively rich in cement—a coarse sand is best. For a lean mortar, a small admixture of fine sand with the coarse is beneficial. If very fine sand is the only kind available, the mixture must be excessively rich in cement, and a smaller quantity of the sand in proportion to the gravel or stone may be used. To secure great density and practical water-tightness of concrete, a larger percentage of fine grains is neded than is theoretically necessary for maximum strength. If a good, coarse sand is not available, almost any size of sand may be employed, provided a sufficient proportion of cement is used; though in this case, tests should be made to see that the mortar or concrete sets and hardens properly. There is one exception, however, to this rule—namely, in the case of concrete to be laid in sea water, where fine sand must never be used.

To choose between two sands, make each into a mortar with cement in the required proportions; and the sand producing a smooth mortar having the smallest volume will give the greatest strength.

Coarseness and **cleanness** are the two most important characteristics of sand. **Sharpness**—ascertained by examination under a magnifying glass, or, more simply, by the gritty "feel" of the sand when rubbed between the fingers—is, sometimes specified as a necessary requirement, but is rather an indication of cleanness and purity, showing no excess of fine material, than an absolute essential of strength.

A sand which soils the hands when rubbed between them, or which has little of the gritty feeling, should not be used. Another method of determining whether the sand is clean is to drop a quantity into a pail of clear water. If the water in two minutes is clear enough to enable you to see the sand at the bottom, the sand is clean. Still another method is to put some of the sand into a fruit jar with some water shake well, and let the jar stand till all the materials have settled. If a layer of mud is seen over the sand, the sand is not clean, and should not be used. A small percentage of dirt, such as fine clay or other mineral matter, is not necessarily harmful; and for lean mortar, the presence of such impurities—up to, say, 10 per cent—may be actually beneficial. For rich mixtures, on the other hand, an excess of dirt or fine material spells weakness, and setting may be seriously delayed.

The presence of any considerable proportion of mica will destroy the value of any sand for concreting purposes. Its effect is more injurious upon fine sands than upon coarse. It greatly increases the amount of voids, and the smooth surfaces of the mica particles do not permit of a good bond with the surrounding cement.

The weight of sand is to some extent an indication of its quality. A heavy sand is generally denser, and therefore better, than a light sand, as it ordinarily indicates coarseness and good grading of particles, with correspondingly low percentage of voids. This, however, is a very unsafe guide to follow, on account of the fact that the presence of moisture may cause as much as 20 per cent variation in weight, and fine sands require more water than coarse. That sand which, with the same percentage of cement, gives the heaviest and densest mixture, is the best.

It is better to err on the side of coarseness than of fineness, for, other things being equal, a coarse sand will give a denser and stronger concrete or mortar; but a judicious proportioning of the fine and the coarse will insure the best results all around; and, in the case of water-tight work, it seems to be the preferred practice to make a slight sacrifice of the theoretical requirements of strength, by increasing the proportion of fine material—except, as already noted, in the case of concrete to be deposited under sea water.

The importance of a careful grading of different-sized aggregates as affecting the strength of concrete, is graphically illustrated in the samples of natural bank sand and gravel shown in Plate 1. The samples are photographed and reproduced at actual size.

In Plate 1, both samples (A and B) were taken from the same natural bank of sand and gravel of glacial origin, at Attica, Indiana. They differ only in the size of the grains, due to screenings, and give, in a general way, a fairly correct idea of the appearance of a well-graded sand. A, the finer sample, contained 34 per cent of voids; B, the coarser, 26.9 per cent. When made up into a 1 : 3 mortar—that is, 1 part cement to 3 parts aggregate—the mortar from the coarser material, in one year, developed a compressive strength of 7,750 lbs. per square inch, whereas the finer material gave a mortar having a compressive strength of only 4,475 lbs. per square inch—a difference in strength of over 70 per cent in favor of the coarser material.

The relative percentage of voids in two materials, one of uniform size and shape, and the other of irregular size and shape, is clearly shown by comparing the gravel of Sample B, Plate 1, with a mass of round shot of uniform size. In the irregular gravel, the voids equal 26.9 per cent, while in the shot they equal 47.6 per cent. That is to say, there will be nearly twice the space in gravel to fill with cement if the particles are round and all the same size.

Selection of Stone and Gravel. So wide a variation in texture and strength is found in stones, even of the same class, that no hard and fast rule can be laid down as to their comparative values for concrete. The actual, intrinsic strength of the **stone itself** is an all-important factor, the hardest stone—other things being equal—producing the strongest concrete.

In order of their value for concreting purposes, different stones may be listed approximately as follows: (1) Trap; (2) **Granite**; (3) **Gravel**; (4) **Marble**; (5) **Limestone**; (6) **Slag**; (7) **Sandstone**; (8) **Slate**; (9) **Shale**; (10) **Cinders.** Sandstone is ordinarily not worth more than three-fourths the value of trap, and the value of slate is less than half that of trap.

Where stone is used as aggregate, the relative size of the particles affects the quality of the concrete, just as in the case of sand; but the strength and density of the concrete depend also on the **hardness** of the stone itself, on the **shape** of the particles, and on their being kept within certain limits as to **maximum size.**

Porous stones, like sandstone, should be avoided if possible. They are inherently weak, and, unless thoroughly soaked before mixing, will give trouble by absorbing excessive amounts of water, thus preventing plasticity and a thorough mixture. The best crushed stone is that which is hard; with angular fragments, preferably of cubical shape and ranging up to the maximum size that can be handled in the work; with all particles smaller than, say, one-fourth inch screened out to be used as sand, and with the sizes of the remaining coarse stone varying from small to large, the coarsest predominating. A flat-grained material packs less closely than, and is always inferior to,. stone of angular or cubical fracture.

Gravel should also meet these same requirements, except as to angular shape, its particles being usually rounded by erosion. It must, in addition, be clean, because clay or loam adhering to the particles weakens or destroys the adhesive power of the cement. Tests made by the Boston Transit Commission showed a tensile strength of 605 lbs. per square inch for concrete made with clean gravel, as against 446 lbs. when made with dirty gravel.

If a gravel and a crushed stone are of equal hardness, the stone, on account of its angularity, will probably give a better binding quality to concrete than will the gravel, which is smooth and rounded. Gravel, however, is used quite as extensively as crushed stone, and with good results; and it has frequently the advantage of greater cheapness and greater accessibility of supply.

Even in the same bank of gravel or the same pile of crushed stone, the proportions of fine and coarse material will vary at different points. As a result, the only way to secure absolute uniformity in the concrete, is to separate the fine material from the coarse by screening, and then remix in the proportions required by the specifications. This adds to the cost, but insures the best results.

Sizes to Use. In the case of both stone and gravel, the coarse material should predominate, and, as already noted, the size of the coarsest particles of stone should be as large as can be handled in the work. Not only is the strength of the concrete increased thereby, but the use of leaner and cheaper mixtures is rendered feasible with equally good results. In mass concrete, however, the stones, if too large, are liable to separate from the mortar unless placed by hand or derrick as in rubble concrete, and a practical maximum size is 2½ or 3 inches. In thin walls, floors, columns, tanks, conduits, and other reinforced constructions, a 1-inch maximum size is generally as large as can be easily worked between the steel, though sometimes 1½-inch is used. In some cases where the walls are very thin—say 3 or 4 inches—a ¾-inch maximum size is more convenient to handle.

Rubble Stone and its Use. Sometimes a saving of expense may be secured in the construction of massive work such as walls, piers, abutments, dams, breakwaters, etc., by the introduction of large stones into the mass. When this is done, the stones should be used only in the middle or center of the wall, and each one should be so placed that there will be ample room between it and its neighbor. Thorough tamping about the stone is necessary to insure the cement adhering to its surface, and also to insure a perfect filling of the voids. Stones used in this way are better if their surface is irregular than if they are smooth or rounded, as the voids of the irregular stone afford better holding power for the cement.

These large, irregular pieces of stone are called rubble, and the concrete made by using them is known as rubble concrete.

In some classes of work, great weight in concrete is desirable—as, for example, in breakwaters, dams, etc., where there will be great shock or pressure; and to such cases rubble concrete is very well adapted, as it is much heavier per cubic foot or cubic yard than ordinary concrete. Compared with rubble masonry, it will usually be cheaper than the latter. This will always be the case if the rubble concrete can be placed without the construction of forms; but in the case of walls of medium thickness, say 3 or 4 feet, the saving in the concreting material will ordinarily be more than offset by the cost of the forms.

In proportioning rubble for concrete, the amount of stone is fixed at a certain percentage of the space to be filled. A greater percentage of a given volume will be filled if the stone is large size than if it is small size. If the rubble is of such size that it can be handled regularly by one man or two men, about 20 to 25 per cent of the space filled by the concrete will consist of the stone, leaving 80 to 75 per cent to be filled with mixed concrete. In the case of stones that can be handled only by a derrick, the percentage may run from about 33 per cent for the smaller sizes to from 55 to 65 per cent for large-size stones averaging 1 to 2 cubic yards each or larger, the amount of mixed concrete necessary thus ranging from 66 to 35 per cent or less, according as the stone is increased in size.

The stones should be completely enveloped in the concrete, and the outermost pieces should be buried at least 5 to 12 inches beneath the surface. A very wet mixture is generally used in placing rubble concrete; and in this case the stones may be placed much closer together than when a medium or dry mixture is used. If the concrete is not mushy enough to flow into and fill all the spaces, it must be thoroughly tamped to insure that none of the crevices are left empty.

Fireproof Aggregates. Aggregates differ greatly in the resistance they offer to high temperatures. But even an aggregate naturally weak in this respect—such as limestone—will give a concrete available for fireproof construction, if, in depositing the concrete, the precaution is taken to insure the burial of the coarser aggregate behind a facing very rich in cement.

Slag and pure quartz sand are two of the best fireproof aggregates. The former is itself a product of enormously high temperatures. Its angularity, density, and cheapness are all in its favor. In the selection of slag, however, care should be taken to avoid any that contains sulphur, since that chemical is detrimental to Portland cement.

Other substances that are considered fire-proof are **firebrick, broken pottery,** hard **clinkers, cinders,** and **pumice-stone** or **lava,** all of which have been through the fire. If cinders are used, they should be thoroughly screened through a mason's screen, to remove the dust.

Limestone, granite, and flint crack and even decompose under the action of great heat. Concrete made from "coke breeze," as the refuse from coke furnaces is called, will gradually crack and crumble when subjected to high temperatures.

WATER FOR CONCRETE

Water has two functions to perform in concrete work. Its presence is necessary to develop the hydraulic activity of the cement. Aside from this, the part it plays in concrete is merely that of rendering the mass plastic and insuring a thorough mixture. It does not seem to have any direct bearing on the adhesive power of cement; but it indirectly affects the strength of concrete through its action in rendering possible a more thorough compacting of ingredients, and consequently increasing the density of the resulting mass of concrete.

In all ordinary concrete work, the general rule may be laid down, that **water should be liberally used.** This greatly improves the density and strength of concrete. Except in special cases—as, for example, in making concrete blocks or casts where the product has almost immediately to be self-supporting—enough water must be used to make the concrete thoroughly soft and plastic, so as to quake strongly when rammed. If mixed too dry, the concrete will never harden properly, and will be light, porous, and crumbling. In some work—as, for example, in placing concrete around reinforcing steel members—the best results can be secured only by an excessively wet mixture, mushy, and soft enough to run off the shovel.

Water that is to be used for concrete should, above all things, be clean. Dirty water will have the same effect as dirty sand or dirty gravel, weakening the concrete and in some cases retarding the setting.

Fresh cement requires more water than cement which is stale. In cold weather, the water may be warmed; but in such cases, there arises the danger of having bubbles of air or of water vapor scattered through the interior of the concrete, leaving small holes in the mass that weaken it.

Water should also be **free from acids** or **strong alkalis**. Cement is chemically basic, and is therefore naturally active to a greater or less extent with the various acids. If concrete is to be exposed to attack from acid-laden fluids, a reliable waterproofing process should be adopted.

The action of **sea water** on Portland cement concrete has exercised the minds of engineers for many years. The tendency to disintegration that has been noticed is purely the result of chemical action and the formation of soluble salts. It is checked by the use of cements relatively low in lime, alumina, and sulphates, and by the adoption of a waterproofing process which will prevent the diffusion through the concrete of the active constituents of the sea water.

Alkalis in water may hasten the setting and hardening of concrete, but are ultimately detrimental to strength.

PROPORTIONING OF CONCRETE INGREDIENTS

The importance of a careful grading of coarse and fine ingredients has already been emphasized, and the general principles laid down which should govern the selection of aggregates for any proposed work.

It is equally essential, for the sake of economy and efficiency, that the concrete worker shall understand how to determine exactly the relative proportions of cement, sand, and coarser aggregates which must be used to obtain the densest and strongest concrete.

At the outset, it should be noted that in practical construction work, proportions are always indicated by measured volume, not by weight. Thus a concrete designated 1:2:4 does not mean one containing 1 lb. cement to 2 lbs. sand to 4 lbs. of coarser aggregate; but it means a mixture consisting of 1 part or measured volume of cement, twice as much sand, and four times as much of the coarser aggregate (stone or gravel), making the whole mixture contain seven parts, though, when mixed together, these seven parts will not occupy seven times the volume of one part, but only a little more than the volume

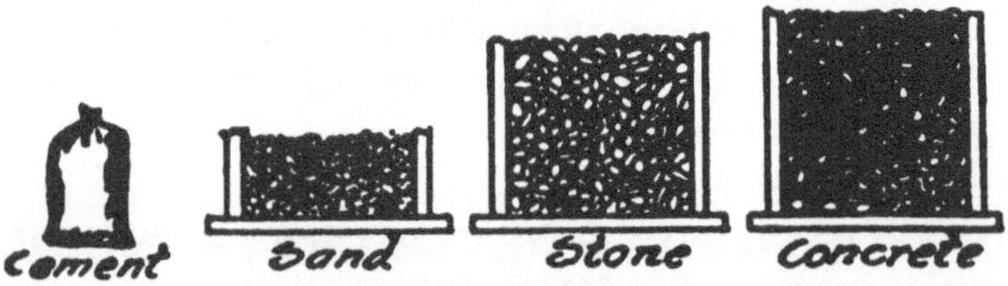

Fig. 2. Relative Volumes of the Solid Ingredients in a 1:2:4 Mixture, and of the Resulting Mass of Concrete.

of the four parts. Fig. 2 illustrates graphically the relative volumes of the ingredients in a 1:2:4 mixture, and of the resulting mass of concrete.

In general practice, there is wide variation in regard to the actual proportions used in the mixing of concrete. In many cases these proportions are decided upon in so crude a manner, and the measuring is also done so roughly and carelessly, that material is wasted, the cost of work unnecessarily increased, and the results indifferent or unreliable.

The only way to determine the **exact** proportions of cement and aggregates required for perfect concrete, is to make an **experimental test** in each case. Fortunately this is a comparatively easy thing to do; and, for the practical concrete worker, it will ordinarily consist merely in determining the **percentage of voids** contained in the sand and other aggregates—that is, the proportion of space that has to be filled with fine material.

As a result of the number of such special tests which have been made by engineers, and the careful investigations that have been carried out in connection with actual examples of construction, an immense fund of valuable practical information has been accumulated, covering every detail of concrete work. This information has been embodied in the form of working **rules** and **labor-saving tables** which render it immediately available for practical application. By merely referring to these tables and observing these rules, the concrete worker can determine with a degree of accuracy sufficient for all practical purposes the proportions of cement and aggregates to be used in any kind of concrete work, and can be sure that he is adopting the proper methods of procedure.

No table, however, can be compiled, from which at least some slight variation is not allowable.

DETERMINATION OF VOIDS

Since the object, in mixing materials of different sizes in concrete, is to fill all the spaces, it follows that the voids in the coarse aggregate will indicate the amount of sand to use, and the voids in the sand will indicate, though but roughly, the amount of cement to use.

Voids are usually referred to as a certain percentage of the whole volume. A gravel, for example, with 35 per cent of voids, means that in a volume of the gravel occupying 100 cubic feet of space, 35 cubic feet consists of voids or empty spaces that have to be filled.

Having selected the available stone or gravel best adapted for the character of construction desired, any one of several different methods may be employed to determine the percentage of voids.

Voids in Coarse Aggregates.—A very simple method of determining voids in coarse aggregates, described by Albert Moyer, is in substance as follows:

TABLE III

Percentage of Voids in Coarse Aggregates

Weight per Cubic Foot	Gravel	Sandstone	Limestone, Medium Soft	Limestone, Medium Hard; Sandstone, Hard	Granite; Blue Stone; Limestone, Hard	Granite, Hard, Trap, Medium	Trap, Hard
70 lbs.	57	53	55	57	58	60	61
75 "	54	50	52	54	55	57	59
80 "	51	47	49	51	52	54	56
85 "	48	43	45	48	50	51	53
90 "	45	40	42	45	47	48	50
95 "	42	37	39	41	44	46	47
100 "	39	33	36	38	41	43	45
105 "	36	30	33	35	38	40	42
110 "	33	26	29	32	35	37	39
115 "	30	23	26	29	32	34	36
120 "	27	20	23	26	29	31	34
125 "	24	16	20	23	26	28	31
130 "	20	13	17	20	23	26	28
135 "	17	10	13	17	20	23	26
140 "	14	6	10	14	17	20	23

Make a box that will contain 3 cubic feet (say, 1 by 1½ by 2 feet, inside dimensions). Dry the aggregate by heating to 212 degrees F., the boiling point of water. If sand or screenings are present, they should be sifted out through a ¼-inch mesh sieve, to be used as sand. Throw stone in box loose, a small quantity at a time, and level off top with straight-edge. Having first weighed the empty box, weigh the box when filled; deduct the weight of the empty box from the gross weight, and divide the net weight (which is the weight of the contents) by 3, which will give the actual weight of one cubic foot. Then simply use Table III, which shows at a glance the percentage of voids in coarse aggregates of varying weight per cubic foot.

A gravel, for example, which on being well shaken down weighs 110 lbs. per cubic foot, is shown by the table to contain 33 per cent of voids.

Another method of determining the voids in coarse aggregates, is given by S. B. Newberry, substantially as follows:

Use a metal box of exactly one cubic foot capacity. Fill with the material to be tested; shake down well; strike off level; and weigh. Now pour water in, at one corner, until it rises even with the surface; and weigh again. The difference in the weights will be the weight of the water that was added to fill the voids. Now, since one cubic foot of water weighs 62.5 lbs., to find the percentage of the volume occupied by the water that just filled the voids, we divide the weight of the latter by 62.5 and multiply by 100.

Suppose, for example, that the box of dry aggregate weighs 92 lbs., and, when the water is poured in, it weighs 120 lbs. Then 120—92=28 lbs., the weight of the water. The amount of the voids will therefore be $(28 \div 62.5) \times 100 = 44.8$ per cent.

If the aggregate is dry and porous, it should be wetted before testing for voids, and then dried just enough to remove all moisture from the surface of the material. In this way the error that might occur through the absorption of water will be eliminated.

The object of pouring in the water at one point only, is to avoid entrapping bubbles of air.

Voids in gravel are seldom less than 30 per cent or more than 45 per cent. Broken stone, being angular, does not compact as readily as gravel, and hence has larger voids.

Voids in Sand or Screenings. To determine the percentage of voids in sand or screenings, Mr. Newberry's method as described above for coarse aggregates, may be used with slight modification. Instead of pouring the water into the sand-filled vessel, the sand should be poured into a vessel containing water. If the water be poured on the sand, it will be impossible to get rid of all the air. The vessel should be filled not less than half-full of water, and the sand introduced slowly. As most sands have less than 50 per cent of voids, the water will overflow the vessel before it is level full of sand. Keep on putting in sand, however, until the vessel is level full, being careful at the same time to have the overflowing water carry off as little as possible of the finer sand. When the vessel is level full of the water-soaked sand, strike off smoothly over the top, and weigh. The calculation of voids is then made in exactly the same manner as for coarse aggregates.

Still another method of determining the voids in sand is to calculate it from the weight of a cubic foot and the **specific gravity** of quartz. The specific gravity of a substance is its relative weight compared with water as the unit. A substance having a specific gravity of 2.5, for example, will weigh 2.5 times as much as water, volume for volume. Now, sand is mostly broken quartz, and quartz has a specific gravity of 2.65. A cubic foot of water weighs 62.5 lbs. Therefore a cubic foot of quartz sand, if free from voids, would weigh 2.65×62.5=165.625 lbs. Now take a cubic foot of the sand to be tested, well shaken down, and weigh it. Suppose, for example, that it weighs 112 lbs. Subtract this from 165.625 lbs., the weight of a solid cubic foot of quartz, and we find a difference of 165.625—112 =53.625 lbs., which is the loss of weight due to the voids. The percentage of voids will therefore be (53.625÷165.625)×100=32.3 per cent.

AMOUNT OF CEMENT TO USE

The amount of cement to be used in mixing concrete is generally stated as a certain **volume or measure**, in proportion to the number of volumes of the sand and the coarser aggregate, thus, 1:2:4 indicating a mixture in which there will

TABLE IV

Proportions of Cement, Sand, and Stone for Different Percentages of Voids in Sand and Stone

Voids in Sand (Per Cent)	Parts of		Voids in Stone (Per Cent)														
	Cement	Sand	36	37	38	39	40	41	42	43	44	45	46	47	48	49	50
			Parts of Stone														
19	1	3.5	7.5	7.5	7.0	7.0	7.0	7.0	6.5	6.5	6.5	6.5	6.0	6.0	6.0	6.0	6.0
20	1	3.5	7.0	7.0	7.0	7.0	6.5	6.5	6.5	6.5	6.0	6.0	6.0	6.0	5.5	5.5	5.5
22	1	3.0	7.0	6.5	6.5	6.5	6.5	6.0	6.0	6.0	6.0	5.5	5.5	5.5	5.5	5.5	5.0
24	1	3.0	6.5	6.0	6.0	6.0	6.0	6.0	5.5	5.5	5.5	5.5	5.5	5.0	5.0	5.0	5.0
26	1	3.0	6.0	6.0	6.0	5.5	5.5	5.5	5.5	5.0	5.0	5.0	5.0	5.0	5.0	4.5	4.5
27	1	2.5	6.0	6.0	5.5	5.5	5.5	5.5	5.0	5.0	5.0	5.0	5.0	4.5	4.5	4.5	4.5
28	1	2.5	5.5	5.5	5.5	5.5	5.5	5.0	5.0	5.0	5.0	5.0	4.5	4.5	4.5	4.5	4.5
29	1	2.5	5.5	5.5	5.5	5.0	5.0	5.0	5.0	5.0	4.5	4.5	4.5	4.5	4.5	4.5	4.5
30	1	2.5	5.5	5.5	5.0	5.0	5.0	5.0	5.0	4.5	4.5	4.5	4.5	4.5	4.5	4.0	4.0
31	1	2.5	5.5	5.0	5.0	5.0	5.0	5.0	4.5	4.5	4.5	4.5	4.5	4.5	4.0	4.0	4.0
32	1	2.5	5.0	5.0	5.0	5.0	5.0	4.5	4.5	4.5	4.5	4.5	4.5	4.5	4.0	4.0	4.0
33	1	2.5	5.0	5.0	5.0	4.5	4.5	4.5	4.5	4.5	4.5	4.0	4.5	4.0	4.0	4.0	4.0
34	1	2.5	5.0	5.0	4.5	4.5	4.5	4.5	4.5	4.5	4.5	4.0	4.0	4.0	4.0	4.0	4.0
35	1	2.0	5.0	4.5	4.5	4.5	4.5	4.5	4.5	4.0	4.0	4.0	4.0	4.0	4.0	4.0	3.5
36	1	2.0	4.5	4.5	4.5	4.5	4.5	4.5	4.0	4.0	4.0	4.0	4.0	4.0	3.5	3.5	3.5
37	1	2.0	4.5	4.5	4.5	4.5	4.0	4.0	4.0	4.0	4.0	4.0	4.0	3.5	3.5	3.5	3.5
38	1	2.0	4.5	4.5	4.5	4.0	4.0	4.0	4.0	4.0	4.0	4.0	3.5	3.5	3.5	3.5	3.5
39	1	2.0	4.5	4.5	4.0	4.0	4.0	4.0	4.0	4.0	4.0	3.5	3.5	3.5	3.5	3.5	3.5
40	1	2.0	4.5	4.0	4.0	4.0	4.0	4.0	4.0	4.0	3.5	3.5	3.5	3.5	3.5	3.5	3.5
41	1	2.0	4.0	4.0	4.0	4.0	4.0	4.0	4.0	3.5	3.5	3.5	3.5	3.5	3.5	3.5	3.5
42	1	2.0	4.0	4.0	4.0	4.0	4.0	4.0	3.5	3.5	3.5	3.5	3.5	3.5	3.5	3.0	3.0

have to be measured out twice as much sand as cement, and four times as much gravel or stone.

The **working rules and tables** which have been compiled by engineers from the results of careful tests and wide observation of actual constructions, afford to all workers in concrete a ready guide to approved practice in the proportioning of mixtures. A number of these practical rules and tables will be given, sufficient to enable the reader to determine the proper mixture to use in any kind of concrete construction.

In the first place, the general principle may be laid down, that economy in the use of cement demands a careful **grading of the aggregates** so as to reduce voids as much as possible. If the coarse aggregate, for example, is very uniform in size, and its particles considerably smaller than the maximum allowed, the voids can be reduced by adding coarser material. If, on the other hand, the aggregate is already almost the full allowable size, the voids can be reduced by adding finer material ranging in size from the coarse aggregate to the coarsest sand. In all aggregates, in fact, the particles should range from fine to coarse, with the coarse predominating, as before explained.

Having selected the sand and coarse aggregate available for the work, and calculated the percentage of voids, the proper mixture to use can be readily determined by reference to Table IV, compiled by Mr. C. W. Boynton, which shows the proportions of cement, sand, and stone to be used for various percentages of voids. This table, in order to be on the safe side and cover all the slight errors likely to occur in determining the voids, is based on an extra allowance of 10 per cent in the voids of the small aggregate. By using this table, a great waste of cement may be prevented.

Where a sand contains 30 per cent voids, for example, the table shows that only 2.5 parts of it should be mixed with 1 part of cement. And if coarser material is also available, say stone containing 45 per cent of voids, then 4½ parts of such aggregate should be added. The proper mixture would be 1:2.5:4.5.

When a natural mixture of sand and gravel is available, the most reliable results will be obtained by screening it, and then remixing the fine and coarse material in the proper definite proportions. But if this cannot be done and unscreened gravel has to be used, the proportion of cement can be roughly determined in the following manner: Take a portion of the gravel, and screen out the sand through a ¼-inch screen. If the volume of sand that passes the screen bears to that retained on the screen the ratio of, say, 3 to 5, then the mixture to use is 1 part cement to 3+5=8 parts of the unscreened gravel, since this practically amounts to a mixture of 1 part cement, 3 parts sand, and 5 parts gravel.

The proportions for a mixture can also be calculated, though but roughly, from the voids in the sand and stone. Assume, for example, a sand with 33 per cent and a gravel with 40 per cent of voids. Then, to fill the voids in 100 volumes of gravel, we shall need 40 volumes of sand. To fill the voids in the sand with cement, we shall need 33 per cent of 40 volumes; but, allowing an extra 10 per cent for thorough coating of the particles and to avoid results of possible errors made in determining the voids, we shall take 43 per cent of 40 volumes, that is, 17.2 volumes of cement. The proportions of the mixture will then be 17.2 cement to 40 sand to 100 gravel, or, approximately, 1:2.3:5.8.

It is a common practice to increase the proportion of mortar more or less above that calculated, to compensate for imperfect mixing or to increase the strength of the concrete. In some cases a rich concrete may be necessary; in others a poorer mixture will do equally well and be far more economical. This will depend on the character and location of the construction. However, proportions carefully determined according to methods such as those described above, are those which, with only slight variations, should be depended upon where very careful work and an economical use of cement combined with great density and strength are required.

It is a widely accepted conclusion among engineers, that a determination of the voids in sand affords at best only an approximate indication of the amount of cement that should be mixed with the sand in making mortar or concrete. Many engineers therefore prefer to base their proportioning of mixtures, so far as the cement and sand are concerned, not on the measured or computed volume of voids in the sand, but on the density and plasticity of the mortar produced by differently proportioned mixtures of cement and sand. The sample mixtures for examination (say five samples in proportions 1:2, 1:2.5, 1:3, etc.) are very carefully proportioned out by weighing, and are made to approximate as nearly as possible the conditions that will exist in actual work, as to compactness, amount of moisture absorbed, etc. The tests are taken in finely graduated measuring tubes which show accurately the relation between the original volume of the sand and that of the resulting mixture, and also the slightest variations in volume of the mixture due to the different proportions. The density of the mortar increases with the amount of the cement, up to that point where a further increase of cement will cause an increase in the volume of the mortar. If the conditions of the work do not require a very dense or strong mortar, the proportions to be adopted will be indicated by one of the samples containing the least cement, but showing sufficient plasticity to insure a good bond in the concrete.

In principle this method of determining proportions is very similar to that depending on the

TABLE V

Proportions of Aggregates Giving Maximum Density with Varying Mixtures of Cement and Sand

Voids in Stone (Per cent.)	Proportions of Mortar									
	1:1	1:2	1:2½	1:3	1:3½	1:4	1:4½	1:5	1:5½	1:6
	Proportions of Stone									
20	5	10	12½	15	17½	20	22½	25	27½	30
21	4¾	9½	12	14¼	16⅔	19	21⅓	23¾	26¼	28½
22	4½	9	11⅓	13⅔	16	18⅕	20½	22¾	25	27¼
23	4⅓	8⅔	10⅞	13	15¼	17⅖	19½	21¾	24	26
24	4	8⅓	10⅖	12½	14⅗	16⅔	18¾	20⅚	23	25
25	4	8	10	12	14	16	18	20	22	24
26	3⅚	7⅔	9⅗	11½	13½	15⅖	17⅓	19¼	21⅕	23
27	3⅔	7⅖	9¼	11	13	14⅘	16⅔	18½	20⅓	22¼
28	3⅗	7⅛	9	10¾	12½	14⅓	16	17⅞	19⅔	21⅖
29	3⅖	7	8⅗	10⅓	12	13⅘	15½	17¼	19	20⅔
30	3⅓	6⅔	8⅓	10	11⅔	13⅓	15	16⅔	18⅓	20
31	3¼	6⅖	8	9⅔	11⅓	13	14½	16	17¾	19⅓
32	3	6¼	7⅘	9⅜	11	12½	14	15⅗	17⅕	18¾
33	3	6	7⅗	9	10⅗	12	13⅔	15⅕	16⅔	18⅕
34	3	6	7⅓	8⅘	10¼	11¾	13¼	14¾	16⅕	17⅔
35	2⅚	5¾	7⅛	8⅗	10	11⅖	12⅚	14⅓	15¾	17⅛
36	2¾	5½	7	8⅓	9¾	11	12½	14	15¼	16⅔
37	2⅔	5⅖	6¾	8	9½	10⅘	12⅕	13½	14⅚	16¼
38	2⅗	5¼	6⅗	8	9¼	10½	11⅚	13⅙	14½	15⅘
39	2½	5	6⅖	7⅔	9	10¼	11½	12⅘	14	15⅓
40	2½	5	6¼	7½	8¾	10	11¼	12½	13¾	15
41	2⅖	4⅞	6	7⅓	8½	9¾	11	12⅕	13⅖	14⅗
42	2⅖	4¾	6	7⅛	8⅓	9½	10¾	12	13	14¼
43	2⅓	4⅔	5⅘	7	8⅛	9⅓	10½	11⅗	12⅘	14
44	2¼	4½	5⅔	6⅘	8	9	10¼	11⅓	12½	13⅔
45	2¼	4⅖	5½	6⅔	7¾	8⅞	10	11	12¼	13⅓
46	2⅙	4⅓	5⅖	6½	7⅗	8⅔	9¾	10⅞	12	13
47	2⅛	4¼	5⅓	6⅖	7½	8½	9⅗	10⅔	11¾	12¾
48	2	4	5⅕	6¼	7¼	8⅓	9⅜	10⅖	11½	12½
49	2	4	5	6⅛	7⅛	8⅙	9⅕	10⅕	11¼	12¼
50	2	4	5	6	7	8	9	10	11	12
51	2	3⅞	5	6	6⅞	7⅚	8⅘	9⅘	10¾	11¾
52	2	3⅚	4⅘	5¾	6¾	7⅔	8⅔	9⅗	10⅗	11½
53	2	3¾	4¾	5⅔	6⅗	7½	8½	9⅖	10⅖	11⅓
54	1⅚	3⅔	4⅗	5½	6½	7⅖	8⅓	9¼	10⅕	11
55	1⅘	3⅔	4½	5½	6⅓	7¼	8⅕	9	10	11
56	1¾	3½	4½	5⅓	6¼	7⅛	8	9	9⅘	10¾
57	1¾	3½	4⅖	5¼	6⅛	7	8	8¾	9⅔	10½
58	1¾	3⅖	4⅓	5⅙	6	7	7¾	8⅗	9½	10⅓
59	1⅔	3⅖	4¼	5	6	6¾	7⅗	8½	9⅓	10⅙
60	1⅔	3⅓	4	5	5⅚	6⅔	7½	8⅓	9⅙	10

"yield" or "volumetric test" for sand, which has already been referred to. Table V is a working table based on this method, compiled by Albert Moyer from the results of many tests. It shows the proportions of aggregates which will give maximum density with a minimum of cement.

The voids in the stone, and the economic proportions of cement to sand which will give the density and plasticity of mortar required on the particular concrete job, should first be determined; and the table may then be applied to show the amount of coarse aggregate to be used.

For example, suppose that a mortar mixture of 1 part cement to 2½ parts sand (1:2½) is decided upon, and that the available coarse aggregate contains 33 per cent of voids. Then, looking at the table, starting at 33 in the void column at the left, and passing over to the right until we come to the proportion column headed 1:2½, we find 7½, which is the number of parts of stone to be added to the 1:2½ mortar mixture.

TESTING OF CEMENT

The most accurate and thorough tests of cement are those which are made in professional testing laboratories. These, however, are not always available; and fortunately they are not always necessary. In large constructions such as public works, and in smaller structures where very important values are at stake, laboratory tests are invariably insisted on. In all ordinary work, however, simpler tests, calling for no very elaborate apparatus and requiring no great degree of technical training, will be quite sufficient for practical purposes.

There is a possibility that now and then some small part at least of the output of a cement plant may fall below the regular standard in quality; and for that reason, on all important work, samples should be taken from every shipment—in fact, a sample from every barrel—and these samples should be tested with sufficient care at least to determine the presence or absence of any radical defects which would render the material unfit for use.

The methods of testing recommended by the Committee on Uniform Tests of Cement, of the American Society of Civil Engineers, are fully described in "Radford's Cyclopedia of Construction," Volume V, pages 104-123.

EASILY-MADE TESTING APPARATUS

Mr. W. Purves Taylor, engineer of the Municipal Testing Laboratories of Philadelphia, Pa., has devised simple methods of testing that in a general way are suited to the requirements of cement users, and for which all the necessary apparatus can easily be made by anyone of ordinary mechanical ability. These simple methods of testing are described as follows:

Soundness. Take about half a pound of cement, place it on a clean surface of metal or glass, and form it into a crater. Into the center of the crater pour about a fifth of its weight of water, and mix thoroughly by hand or with a trowel for a couple of minutes, until a stiff and uniform paste is obtained. Make a ball of the paste about 2 inches in diameter, and drop it on the table from a height of about 2 feet. If the ball flattens more than half its depth, the paste is too wet; if it cracks badly, it is too dry. If necessary, add more cement or more water, mix thoroughly, and test until the right consistency is obtained; then mould two balls about 2 inches in diameter.

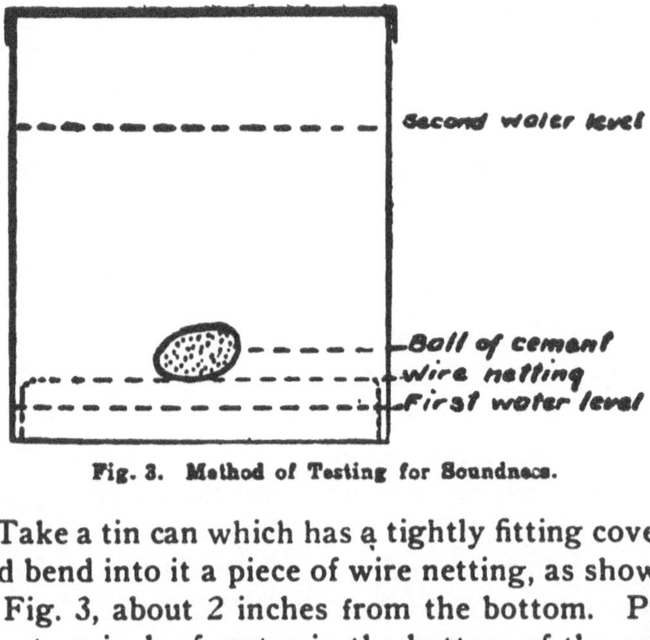

Fig. 3. Method of Testing for Soundness.

Take a tin can which has a tightly fitting cover, and bend into it a piece of wire netting, as shown in Fig. 3, about 2 inches from the bottom. Put about an inch of water in the bottom of the can, place one of the balls on the netting, cover the can, and put it away in a cool place for 24 hours. Then fill the can with water to cover the ball, remove the cover, and place the can on a stove at such a heat that the water will boil in about half an hour. Let it boil gently then for three hours. Have a recond vessel of boiling water beside the can; and, as the water evaporates from the first vessel, replace it with the boiling water from the second one. **Never add cold water.** After boiling three hours, remove the ball, and examine it.

A good Portland cement will always pass this test, and the ball will remain sound and hard. If the ball is disintegrated, or if it is checked or cracked, it generally indicates inferior quality and untrustworthy material. Sometimes failure in boiling is caused by the material being too fresh, and, on a second test made a month or so later, the cement will pass, showing that the expansive elements have become hydrated and thus inert. In general, however, it is on the safe side never to use cement that fails in boiling. If it fails at first, store it away for a month, and then test again. If it still fails, it is better not to use it.

Time of Setting. Put the second ball, made as previously described, in a place protected from the sun or any other source of heat, and from any strong current of air. At the end of twenty minutes, examine it; then put it away, and examine it again in ten hours. The ball, at twenty minutes, should still be soft and pliable, damp on the surface, and should not feel warm. At ten hours it should be dry, firm, and hard enough so that a firm pressure of the thumb nail will make no impression on it. If the cement begins to harden or feel warm in less than twenty minutes, it is generally inadvisable to use it, since setting will have begun before the mortar or concrete is moulded, and the result will be a weak and easily disintegrated product. It is, of course, possible to retemper such cement and obtain excellent results; but this process requires considerable skill and experience, and therefore is not usually to be recommended; moreover, it always means at least some slight loss in strength.

Quick-setting cement often becomes slower by storage for a month or two, so that it may be better to keep such material for that length of time than to attempt to use it earlier at the risk of poor results. A cement that does not harden in ten hours may ultimately give good results; but the slow setting will much delay the progress of the work, and may cause injury in removing the moulds or striking the centers. Safe practice will require cement to set between the stated limits.

Purity. Take as much cement as may be lifted on a five-cent piece, and place it in a china or glass dish. Pour on it a mixture of 1 part of water and 3 parts of muriatic (hydrochloric) acid, using a quantity equal to about three times the volume of cement. Pure Portland cement effervesces violently for a second or two, and then forms a yellowish jelly. A continued effervescence shows adulteration with limestone or natural cement. Cinders or sand, if present, will be shown in a sediment at the bottom of the jelly. The presence of slag is shown by the characteristic putrid odor of hydrogen sulphide (sulphuretted hydrogen). Cement containing any of these adulterations or impurities should not be used.

Strength. It is hard to determine the strength of a cement mortar by any simple method, and

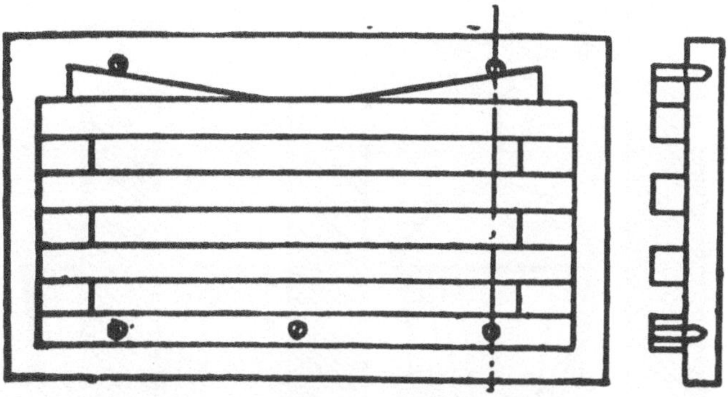

Fig. 5. Moulds for Simple Tests.

obtain results of even approximate accuracy. The following, however, is probably the best simple method.

Make a mould, as shown in Fig. 4, of a planed board and some one-inch strips of wood. Fasten the top strip to the board, and place the others on loosely, in the manner indicated in the diagram, holding them in place by the wedges at the bottom, thus forming moulds for three prisms or bars of mortar 1 inch by 1 inch by 12 inches long. Take special care that the cross-sections of the prisms are exactly one square inch.

Take mortar from the mixing box, or make some of 1 part cement and 3 parts sand, thoroughly mixed dry and then wetted to form a stiff mortar; and fill the moulds, pressing the material in firmly, and smoothing the tops with to prevent mortar from sticking. Put the a trowel. The moulds should be oiled slightly

Fig. 5. Method of Testing Strength.

moulds away in a cool and, if possible, a damp place, for 24 hours. Then remove the prisms of mortar, and place them in fresh water, neither hot nor cold, for 6 or for 27 days. Any prisms that are chipped or otherwise defective should be discarded. Make the apparatus shown in Fig. 5 with a board, a round piece of wood about an inch in diameter, and a pail. The bottom bearing edges are made by cutting the round piece of wood in half, and are fastened to the board exactly 10 inches apart, center to center. The center bearing is a round piece of wood from which a pail is suspended by a cord.

When the test pieces are 7 or 28 days old, place them, still wet, on the apparatus as shown in the cut, taking care that the upper strip is exactly in the center and at right angles with the prism. Adjust the pail so that it is just off the ground and exactly under the specimen; and then slowly pour sand into it until the prism breaks. Carefully weigh the bucket, with the cord and strip still attached; this will indicate the load that stressed the bar to the failing point. If the prism is not exactly one inch square, correct the weight by dividing it by the product of its width times the square of its depth in inches.

The approximate strength of the mortar may then be computed as follows:

Tensile strength equals center load multiplied by 10.

Compressive strength equals center load multiplied by 50.

If the average of three prisms tested in this manner shows a tensile strength of less than 120 pounds at 7 days, or 180 pounds at 28 days, on 1:3 mixture, something is radically wrong with either the cement or the sand, and the fault should be thoroughly investigated.

Tests made by this method are, of course, far from being accurate, but nevertheless give a very fair indication of the value of the material.

In determining the relative value of different sands intended for use, they will often give most valuable and useful information.

Crude as these tests may be said to be, they are vastly superior to no tests at all, and should be of much assistance to the small user in securing good material. Insist on cement that will boil and set normally, and most of the dangers of inferior cement will be avoided.

Another simple method of testing by which the user may ascertain the soundness of the cement he proposes to use, is as follows:

Make three cakes of cement, thicker in the middle than on the edges. Allow one to remain in moist air for 24 hours, and then steam it 4 hours. The second cake should be exposed in moist air, and the third be immersed in water. The results in the case of the second and third cakes should be noted at intervals during 28 days. If the cement is sound, it will not disintegrate; but if it shows expansion cracks on the edges of any of the cakes, it is not sound.

STANDARD SPECIFICATIONS FOR CEMENT

The requirements which a cement must meet in order to be recognized as of "standard" quality, are outlined by the Committee on Standard Specifications for Cement, of the American Society for Testing Materials, in substance as follows:

General Conditions. All cement shall be inspected—either at the place of manufacture or on the work.

In order to allow ample time for inspecting and testing, the cement should be stored in a suitable weather-tight building having the floor properly blocked or raised from the ground, and should be easy of access for proper inspection and identification of each shipment.

Every facility shall be provided by the contractor, and a period of at least twelve days allowed for the inspection and necessary tests.

Cement shall be delivered in suitable packages, with the brand and name of manufacturer plainly marked thereon.

A bag of cement shall contain 94 pounds of cement, net. Each barrel of Portland cement shall contain 4 bags, and each barrel of natural cement shall contain 3 bags of the above net weight.

Cement failing to meet the 7-day requirements may be held awaiting the results of the 28-day tests before rejection.

All tests shall be made in accordance with the methods proposed by the Committee on Uniform Tests of Cement, of the American Society of Civil Engineers (see above).

The acceptance or rejection shall be based on the following requirements:

Specifications for Natural Cement

Fineness. A natural cement shall leave by weight a residue of not more than 10 per cent on the No. 100, and 30 per cent on the No. 200 sieve.

Time of Setting. It shall not develop initial set in less than 10 minutes, and shall not develop hard set in less than 30 minutes, or in more than 3 hours.

Tensile Strength. The minimum requirements for tensile strength for briquettes one inch square in cross-section, shall be within the following limits, and shall show no retrogression in strength within the periods specified:

Age	Strength
NEAT CEMENT	
24 hours in moist air	50-100 lbs.
7 days (1 day in moist air, 6 days in water)	100-200 "
28 days (1 day in moist air, 27 days in water)	200-300 "
ONE PART CEMENT, THREE PARTS STANDARD SAND	
7 days (1 day in moist air, 6 days in water)	25- 75 lbs.
28 days (1 day in moist air, 27 days in water)	75-150 "

For example, the minimum requirement for the 24-hour neat cement test should be some specified value within the limits of 50 and 100 pounds; and so on for each period stated.

If the minimum strength is not specified, the mean of the above values shall be taken as the minimum strength required.

Constancy of Volume. Pats of neat cement about three inches in diameter, one-half inch thick at center, tapering to a thin edge, shall be kept in moist air for a period of 24 hours. One pat shall then be kept in air at normal temperature; and another kept in water maintained as near 70 degrees F. as practicable.

These pats are observed at intervals for at least 28 days, and, to satisfactorily pass the tests, should remain firm and hard, and show no signs of distortion, checking, cracking, or disintegration.

Specifications for Portland Cement

Specific Gravity. The specific gravity of the cement, ignited at a low red heat, shall be not less than 3.10, and the cement shall not show a loss in weight, on ignition, of more than 4 per cent.

To determine the loss on ignition, place half a gram of cement in a platinum crucible, set in a hole in an asbestos board so that about three-fifths of the crucible projects below the board. Expose the crucible to blast flame for 15 minutes, preferably with an inclined flame. The loss, found on weighing, should be checked by a second blasting of 5 minutes.

Fineness. The cement shall leave by weight a residue of not more than 8 per cent on the No. 100, and not more than 25 per cent on the No. 200 sieve.

Time of Setting. It shall not develop initial set in less than 30 minutes, and must develop hard set in not less than 1 hour, nor more than 10 hours.

Tensile Strength. The minimum requirements for tensile strength for briquettes one inch square in section, shall be within the following limits, and shall show no retrogression in strength within the periods specified:

NEAT CEMENT.

Age	Strength
24 hours in moist air	150-200 lbs.
7 days (1 day in moist air, 6 days in water)	450-550 "
28 days (1 day in moist air, 27 days in water)	550-650 "

ONE PART CEMENT, THREE PARTS SAND

Age	Strength
7 days (1 day in moist air, 6 days in water)	150-200 lbs.
28 days (1 day in moist air, 27 days in water)	200-300 "

For example, the minimum requirement for the 24-hour neat cement test should be some specified value within the limits of 150 and 200 pounds; and so on for each period stated.

If the minimum strength is not specified, the mean of the above values shall be taken as the minimum strength required.

Constancy of Volume. Pats of neat cement about three inches in diameter, one-half inch thick at the center, and tapering to a thin edge, shall be kept in moist air for a period of 24 hours. One pat shall then be kept in air at normal temperature, and observed at intervals for at least 28 days. Another pat shall be kept in water maintained as near 70 degrees F. as practicable, and observed at intervals for at least 28 days. A third pat shall be exposed in any convenient way in an atmosphere of steam, above boiling water, in a loosely closed vessel, for 5 hours.

These pats, to satisfactorily pass the requirements, shall remain firm and hard, and show no signs of distortion, checking, cracking, or disintegration.

Sulphuric Acid and Magnesia. The cement shall not contain more than 1.75 per cent of anhydrous sulphuric acid (SO_3), nor more than 4 per cent of magnesia (MgO).

THE MIXING OF CONCRETE

Having learned how to select good materials and how to proportion them so as to obtain concrete of any desired strength, the next important requirement for the concrete worker is that he shall know how to decide upon the proper mixture to use for any particular kind of work.

CHOICE OF CONCRETE MIXTURES

Mixtures of greatly different strength and cost, as already explained, can be made from the same materials, by simply combining them in different proportions. The particular mixture to use will depend principally upon the **kind of service** that will be required of the concrete—upon whether it will have to bear heavy or light loads and withstand great or small stresses; whether or not it will be subjected to hard and rough wear; whether it is to be covered or left exposed, above ground or underground; whether it will have to withstand water pressure, etc.

Kinds of Mixtures

Concrete mixtures are classified in two different ways—(1) as to consistency, or relative wetness; and (2) as to richness, or relative quantity of cement.

Wetness of Mixture. Classified according to relative wetness, there are three kinds of mixture commonly used in concrete work—namely, **Very Wet, Medium,** and **Dry.** Their distinguishing features are as follows:

Very Wet Mixture. Concrete wet enough to be mushy and to run off a shovel in handling. Used for reinforced work, such as thin walls, floors, or other thin sections, reinforced columns, tanks, conduits, etc. Practically no ramming is necessary, but spading should be done next to the forms, to insure burial of the coarser aggregate and the formation of an unbroken face by the cement grout.

Medium Mixture. Concrete just wet enough to make it jelly-like, so that it will quake on tamping. Used for some reinforced work, but chiefly for foundations, floors, and massive constructions, such as retaining walls, piers, abutments, arches, etc. Ramming with tamper or treading with feet necessary to remove air-bubbles and fill voids. A man stepping on a pile of this concrete would sink ankle-deep. Spading also necessary for same reason as in the very wet mixture.

Dry Mixture. Concrete like damp earth. Used

for foundations, etc., where it is important that the concrete shall set as quickly as possible. Should be placed in layers 4 to 6 inches deep at a time, and thoroughly tamped until the water comes to the surface. In dry concrete, it is only by thorough tamping that the danger of the formation of pockets which would weaken the concrete can be avoided. Careful spading should also be done next the forms.

The drier the mixture, the sooner the concrete will set. Any quantity of water that falls below the minimum necessary for thorough hydration of the cement, is, of course, not to be allowed in any case, for then the concrete would be weak and crumbly; but beyond this point the addition of more or less water does not seem to have any very important bearing upon the ultimate strength of the concrete, **provided the mixing is thoroughly done and the concrete properly deposited.** A liberal use of water, however, besides increasing plasticity and ease of working, renders easier the insuring of a thorough mixture and the securing of great density. In reinforced concrete work, a dry mixture is not at all satisfactory, since it is more difficult, and hence more costly, to handle than a wet mixture, will not penetrate and fill all spaces between the reinforcing steel members so as to insure the thorough bond that is required, must be protected with greater care from the sun or from drying too quickly, and, unless great skill is used in spading it, will show voids or pockets in the face of the work when the forms are removed. The prevailing tendency in present-day practice, therefore, is toward the medium or very wet mixtures.

Richness of Mixture. When classified according to relative richness in cement, concrete mixtures are distinguished under four different headings, namely, **Rich, Medium,** Ordinary, and **Lean.**

Rich Mixture. A 1:2:4 concrete, that is, a mixture containing ingredients in the proportion of 1 barrel (4 bags) of packed Portland cement just as it comes from the manufacturer, to 2 barrels (7.6 cubic feet) of loose sand, to 4 barrels (15.2 cubic feet) of loose gravel or broken stone —would be called a rich mixture. This gives a very strong concrete adapted to all kinds of reinforced work that will be subjected to very heavy loads and stresses, such as reinforced floors, arches, beams, girders, columns, foundations for heavy vibrating machinery, etc. It is also used where water-tightness and air-tightness are desired, as in the construction of reservoirs, water conduits, silos, tanks, cisterns, troughs, culverts, building walls, etc.

Medium Mixture. A 1:2.5:5 concrete—that is, in the proportions of 1 barrel (4 bags) Portland cement to 2.5 barrels (9.5 cubic feet) loose sand, to 5 barrels (19 cubic feet) loose gravel or broken stone—would be called a medium mixture. This also gives a very strong concrete, and is adaptable to ordinary machine foundations, thin foundation walls, building walls, arches, ordinary ground floors, sidewalks, gutters, sewers, culverts, etc.

Ordinary Mixture. A 1:3:6 concrete—that is, in the proportions of 1 barrel (4 bags) Portland cement, to 3 barrels (11.4 cubic feet) loose sand, to 6 barrels (22.8 cubic feet) loose gravel or broken stone—would be called an **ordinary mixture.** This gives a strong concrete, but of less strength than either a medium or a rich mixture. It is adaptable to massive work, such as heavy walls, retaining walls, piers, abutments, etc., which will be subjected to considerable strain. It is sometimes used for reinforced concrete in floors and walls; but is more generally used without reinforcement, in foundations, footings, ground floors, sidewalks, gutters, etc.

Lean Mixture. A 1:4:8 concrete—that is, in the proportions of 1 barrel (4 bags) Portland cement, to 4 barrels (15.2 cubic feet) loose sand, to 8 barrels (30.4 cubic feet) loose gravel or broken stone—would be called a **lean mixture.** This gives a concrete weak in tensile strength, but of considerable compressive strength, and therefore specially adaptable to comparatively unimportant massive work where the concrete will be subjected only to plain compressive strains, as in large and thick foundations supporting stationary loads, backing for stone masonry, sub-bases for sidewalks, driveways, etc.

TABLE VI

Approximate Mixtures Adaptable to Various Classes of Work

Rich, 1:2:4; Medium, 1: 2.5:5; Ordinary, 1:3:6; Lean, 1:4:8
(A. S. Johnson)

Kind of Work	Mixture	Consistency
Abutments	Rich to Ordinary	Medium
Arches	Rich to Medium	Medium
Backing for Masonry	Lean	Medium to Dry
Beams, Reinforced	Rich to Medium	Very Wet
" Plain	Rich to Medium	Very Wet to Medium
Cisterns	Rich to Medium	Very Wet to Medium
Columns, Reinforced	Rich	Very Wet
Conduits, Water	Rich	Very Wet
Coping	Rich to Medium	Medium
Culverts, Reinforced	Medium to Ordinary	Medium
" Plain	Medium to Ordinary	Medium
Driveways	Same as Sidewalks	
Fence Posts	Rich	Very Wet to Medium
Floors, Reinforced	Rich to Ordinary	Very Wet to Medium
" Ordinary Ground	Medium to Ordinary	Medium
Footings	Ordinary to Lean	Medium
Foundations, Heavy Vibrating Machinery	Rich	Very Wet to Medium
" Ordinary Machinery	Medium	Medium
" Thin Walls	Rich to Medium	Very Wet to Medium
" Thick Walls	Medium to Lean	Medium to Dry
Girders, Reinforced	Rich to Medium	Very Wet
" Plain	Same as Beams	
Gutters	Same as Sidewalks	
Pavements	Same as Sidewalks	
Piers	Rich to Ordinary	Medium
Reservoirs	Rich to Medium	Very Wet to Medium
Roof Slabs	Medium to Ordinary	Medium
Sewers, Reinforced	Rich to Medium	Medium
" Plain	Medium	Medium
Sidewalks (Base)	Medium to Ordinary	Medium to Dry
" (Sub-Base)	Ordinary to Lean	Medium to Dry
Silos	Rich to Medium	Very Wet to Medium
Tanks	Rich to Medium	Very Wet to Medium
Walls, Dwelling Houses	Rich to Medium	Very Wet to Medium
" Large Buildings (Compression and Tension)	Rich to Medium	Very Wet to Medium
" Large Buildings (Compression Only)	Medium to Ordinary	Medium
" Massive	Medium to Ordinary	Medium
" Retaining	Medium to Ordinary	Medium
" Thin Foundations	Rich to Medium	Very Wet to Medium
" Tunnel	Medium to Ordinary	Medium

As a rough guide that will enable the concrete worker or contractor to tell at a glance approximately the kind of mixture appropriate for various classes of work, Table VI has been compiled. The indications of mixtures given in the table must not be taken as exact and final specifications for any one of the particular kinds of work mentioned, but merely as approximate guides to what is average current practice.

Variations Allowable in Mixtures. If, for example, the aggregates are well graded, and their proportions carefully determined by one of the methods before described, smaller proportions of cement may be used for each class of work. If, on the other hand, the sand is very fine, the quantity of cement should be increased 10 to 15 per cent. Instances of construction will occur in which the strength of concrete required for safety may be obtained from almost any mixture ranging from lean to rich; and in such cases, choice of mixture will probably be dictated by considerations of cost or by empirical rules based on what has been found most advantageous in actual practice.

Measuring Out Quantities for Mixing

Some authorities, for the sake of simplicity in proportioning and measuring, consider that 1 bag Portland cement=1 cubic foot=100 lbs., thus making 1 barrel=400 lbs. It is, however, a closer approach to actual facts as found in practice, to take the average weight of a barrel as 380 lbs., thus making the weight of a bag 95 lbs.

Throughout the present work, unless otherwise expressly stated, all proportions will be given **by volume, not by weight.**

A convenient form of measuring box can be made from any kind of rough boards. It may be rectangular in shape with perpendicular sides, and should be provided with a short handle at each corner (see Plate 2). It should have no top or bottom, being filled only at the place of mixing, and emptied by being simply lifted up. It should be of three or four cubic feet capacity, and may be graduated so that one, two, or more cubic feet can be accurately measured.

In measuring, do not pack or ram the material. Sand, gravel, and stone should be measured loose. Simply shovel in the loose material, strike off level across the top, and lift the box away.

Tables for Measuring Materials

In measuring out the quantities required for mixing concrete, it may be necessary to proceed in different ways. We may, for example, have to mix in small batches, and shall need to know how much sand and gravel to lay out for two or three bags of cement at a time; or we may have to decide on the quantities as required per cubic yard of the space to be filled with concrete; etc.

Several tables will therefore be given, meeting these different conditions; and by reference to these tables the quantities required for any ordinary mixture can be determined almost at a glance.

Quantities for Batch Mixing. Tables VII and VIII, compiled by Percy H. Wilson and

TABLE VII

Quantities of Materials, and Resulting Amount of Concrete, for a Two-Bag Batch

Aggregates—Sand and Broken Stone or Screened Gravel

Mixture	Quantities of Material Required			Amount of Concrete Obtained	Size of Measuring Boxes (Inside Dimensions)		Water for Medium Wet Mixture
	Cement	Sand	Stone or Gravel	Cu. Ft.	Sand	Stone or Gravel	
	Bags	Cu. Ft.	Cu. Ft.				Gallons
1:2:4	2	3⅘	7½	8½	2' x 2' x 11½"	2' x 4' x 11½"	10
1:2:6	2	5⅘	11½	12	3' x 3' x 11½"	3' x 4' x 11½"	13½

Clifford W. Gaylord, show the quantities for a 2-bag batch. A 3-bag batch will require half as much more of each ingredient; a 4-bag batch, just double the amount; and so on. If the aggregates used are sand and crushed stone or gravel, use Table VII; if natural-bank sand and gravel already mixed, use Table VIII.

Measure the water with a bucket. This will give more uniform results than using a hose. The number of gallons indicated in the tables is only approximate. Try it on the first batch; and if the mixture is made too wet, reduce the amount of water; if too dry, increase the amount, so as to get the desired consistency.

Suppose, for example, that we wish to construct a silo. This calls for a certain amount of rich mixture 1:2:4 (see Table VI). If a natural mixture of sand and gravel is not obtainable, we shall have to mix these ingredients ourselves, possibly using crushed stone instead of

TABLE VIII

Quantities of Materials, and Resulting Amount of Concrete, for a Two-Bag Batch

Aggregate—A Natural Mixture of Sand and Gravel

Mixture	Proportions by Parts		Quantities of Material Required		Amount of Concrete Obtained	Size of Measuring Boxes (Inside Dimensions)	Water for Medium Wet Mixture
	Cement	Natural Mixture of Sand and Gravel	Cement	Natural Mixture of Sand and Gravel			
			Bags	Cu. Ft.	Cu. Ft.		Gallons
1:2:4	1	4	2	7½	8½	2' x 4' x 11¼"	10
1:3:6	1	6	2	11¼	12	2' x 4' x 11¼"	13½

gravel, and shall therefore have to use Table VII. If, on the other hand, a natural bank of sand and gravel is near by, we shall take our aggregate from that, and in that case shall use Table VIII.

Assume the latter case. We shall probably have first to screen out the excess of fine material so as to get a well-graded aggregate, for natural mixtures ordinarily contain far too much sand or finer material. Then, simply turning to Table VIII, we look for 1:2:4 under the heading "Mixture." Reading across toward the right, we find under the heading "Quantities of Material Required," that we shall have to measure out 7½ cubic feet of the natural mixture of sand and gravel for every 2 bags of cement used. The next column shows that every such batch will give us 8½ cubic feet of concrete; and under "Size of Measuring Boxes" we learn that the mixed sand and gravel for a batch should just fill a box 2 feet wide, 4 feet long, and 11½ inches deep, inside dimensions. Ten gallons of water may be tried in mixing the first batch, but the quantity may have to be increased. This will depend somewhat on the way the aggregate is graded, and the amount of fine material present.

Possibly we may have figured out the number of cubic feet of concrete that will be required on a job, and our cement will be in barrels in-

TABLE IX

Ingredients in One Cubic Foot of Concrete

Mixture	Quantities of Materials		
	Cement	Sand	Stone or Gravel
1:2:4	.058 bbl.	.0163 cu. yd.	.0326 cu. yd.
1:3:6	.041 "	.0174 "	.0348 "

stead of bags. In that case, Table IX will be found useful, as it shows the quantities of cement and aggregates in one cubic foot of concrete of varying mixtures.

For example, suppose the work consists of a concrete silo requiring in all 935 cu. ft. of concrete, of which 750 cu. ft. is to be 1:2:4 concrete, and 185 cu. ft. is to be 1:3:6 concrete. Also enough sand and cement is needed to paint the silo inside and outside, in all 400 sq. yds. of surface, with a 1:1 mixture of sand and cement. One cu. ft. of 1:1 mortar will paint about 15 sq.

yds. of surface, and requires 0.1856 barrel of cement and 0.0263 cu. yd. of sand. Then we have:

Cement—

750 cu. ft. of 1:2:4 concrete is	$750 \times .058$	= 43.5 bbls.
185 cu. ft. of 1:3:6 concrete is	$185 \times .041$	= 7.6 "
Painting is	$\frac{400}{15} \times .1856$	= 4.9 "
Total cement.		56.0 bbls.

Sand—

750 cu. ft. of 1:2:4 concrete is	$750 \times .0163$	= 12.25 cu. yds.
185 cu. ft. of 1:3:6 concrete is	$185 \times .0174$	= 3.25 " "
Painting is	$\frac{400}{15} \times .0263$	= .75 " "
Total sand.		16.25 cu. yds.

Stone or Gravel—

750 cu. ft. of 1:2:4 concrete is	$750 \times .0326$	= 24.5 cu. yds.
185 cu. ft. of 1:3:6 concrete is	$185 \times .0348$	= 6.5 " "
Total stone or gravel.		31.0 cu. yds.

Thus the necessary quantities of materials are 56 bbls. of Portland cement; 16¼ cu. yds. of sand; 31 cu. yds. of stone or gravel.

TABLE X

Materials for One Cubic Yard of Concrete

Proportions	Bbls. Cement	Bbls. Sand	Bbls. Gravel or Stone
1:2:4	1.57	3.14	6.28
1:2.5:5	1.29	3.23	6.45
1:3:6	1.10	3.30	6.60
1:4:8	0.85	3.40	6.80

Quantities per Cubic Yard. Table X, showing the number of barrels of cement, sand, and gravel required for one cubic yard of concrete of different mixtures, will serve for estimating roughly the quantities and cost of materials. A more nearly accurate determination can be made by referring to Table XI, which indicates the number of barrels of Portland cement, and the fraction of a cubic yard of both sand and stone,

TABLE XI

Amount of Material Required for One Cubic Yard Rammed Concrete of Varying Mixture

Mixture			Stone, 1-inch and Under, Dust Screened out			Stone, 2½-inch and Under, Dust Screened out			Stone, 2½-inch, with most Small Stone Screened out			Gravel, ¾-inch and Under		
			Cement, Bbls.	Sand, Cu. Yds.	Stone, Cu. Yds.	Cement, Bbls.	Sand, Cu. Yds.	Stone, Cu. Yds.	Cement, Bbls.	Sand, Cu. Yds.	Stone, Cu. Yds.	Cement, Bbls.	Sand, Cu. Yds.	Gravel, Cu. Yds.
1	1.0	2.0	2.57	0.39	0.78	2.63	0.40	0.80	2.72	0.41	0.83	2.30	0.35	0.74
1	1.0	2.5	2.29	0.35	0.70	2.34	0.36	0.89	2.41	0.37	0.92	2.10	0.32	0.80
1	1.0	3.0	2.06	0.31	0.94	2.10	0.32	0.96	2.16	0.33	0.98	1.89	0.29	0.86
1	1.0	3.5	1.84	0.28	0.98	1.88	0.29	1.00	1.88	0.29	1.05	1.71	0.26	0.91
1	1.5	2.5	2.06	0.47	0.78	2.09	0.48	0.80	2.16	0.49	0.82	1.83	0.42	0.73
1	1.5	3.0	1.85	0.42	0.84	1.90	0.43	0.87	1.96	0.45	0.90	1.71	0.39	0.78
1	1.5	3.5	1.72	0.39	0.91	1.74	0.40	0.93	1.79	0.41	0.96	1.57	0.36	0.83
1	1.5	4.0	1.57	0.36	0.96	1.61	0.37	0.98	1.64	0.38	1.00	1.46	0.33	0.88
1	1.5	4.5	1.43	0.33	0.98	1.46	0.33	1.00	1.51	0.35	1.06	1.34	0.31	0.91
1	2.0	3.0	1.70	0.52	0.77	1.73	0.53	0.79	1.75	0.54	0.81	1.54	0.47	0.73
1	2.0	3.5	1.57	0.48	0.83	1.61	0.49	0.85	1.66	0.50	0.88	1.44	0.44	0.77
1	2.0	4.0	1.46	0.44	0.89	1.48	0.45	0.90	1.53	0.47	0.93	1.34	0.41	0.81
1	2.0	4.5	1.36	0.42	0.93	1.38	0.42	0.95	1.43	0.43	0.98	1.26	0.38	0.86
1	2.0	5.0	1.27	0.39	0.97	1.29	0.39	0.98	1.33	0.39	1.03	1.17	0.36	0.89
1	2.5	3.5	1.45	0.55	0.77	1.48	0.56	0.79	1.51	0.58	0.81	1.32	0.50	0.70
1	2.5	4.0	1.35	0.52	0.82	1.38	0.53	0.84	1.42	0.54	0.87	1.24	0.47	0.75
1	2.5	4.5	1.27	0.18	0.87	1.29	0.49	0.88	1.33	0.51	0.91	1.16	0.44	0.80
1	2.5	5.0	1.19	0.46	0.91	1.21	0.46	0.92	1.26	0.48	0.96	1.10	0.42	0.83
1	2.5	5.5	1.13	0.43	0.94	1.15	0.44	0.96	1.18	0.44	0.99	1.03	0.39	0.86
1	2.5	6.0	1.07	0.41	0.97	1.07	0.41	0.98	1.10	0.41	1.03	0.98	0.37	0.89
1	3.0	4.0	1.26	0.58	0.77	1.28	0.58	0.75	1.32	0.60	0.80	1.15	0.53	0.72
1	3.0	4.5	1.18	0.54	0.81	1.20	0.55	0.82	1.24	0.57	0.85	1.09	0.50	0.75
1	3.0	5.0	1.11	0.51	0.85	1.14	0.52	0.87	1.17	0.54	0.89	1.03	0.47	0.78
1	3.0	5.5	1.06	0.48	0.89	1.07	0.49	0.90	1.11	0.51	0.93	0.97	0.44	0.81
1	3.0	6.0	1.01	0.46	0.92	1.02	0.47	0.93	1.06	0.48	0.97	0.92	0.42	0.84
1	3.0	6.5	0.96	0.44	0.95	0.98	0.44	0.96	1.00	0.45	1.01	0.88	0.40	0.87
1	3.0	7.0	0.91	0.42	0.97	0.93	0.42	0.98	0.94	0.42	1.05	0.84	0.38	0.89
1	3.5	5.0	1.06	0.56	0.80	1.07	0.57	0.82	1.11	0.59	0.85	0.96	0.50	0.76
1	3.5	5.5	1.00	0.53	0.84	1.02	0.54	0.85	1.06	0.56	0.89	0.92	0.48	0.78
1	3.5	6.0	0.95	0.50	0.87	0.97	0.51	0.89	1.00	0.53	0.92	0.88	0.46	0.80
1	3.5	6.5	0.92	0.49	0.91	0.93	0.49	0.92	0.96	0.51	0.95	0.83	0.44	0.82
1	3.5	7.0	0.87	0.47	0.93	0.89	0.47	0.95	0.91	0.49	0.98	0.80	0.43	0.85
1	3.5	7.5	0.85	0.45	0.96	0.86	0.45	0.98	0.86	0.47	1.01	0.76	0.41	0.87
1	3.5	8.0	0.80	0.42	0.97	0.82	0.43	1.01	0.81	0.45	1.04	0.73	0.39	0.89
1	4.0	6.0	0.90	0.55	0.82	0.92	0.56	0.84	0.95	0.58	0.87	0.83	0.51	0.77
1	4.0	6.5	0.87	0.53	0.85	0.88	0.53	0.87	0.91	0.55	0.90	0.80	0.49	0.79
1	4.0	7.0	0.83	0.51	0.89	0.84	0.51	0.90	0.87	0.53	0.93	0.77	0.47	0.81
1	4.0	7.5	0.80	0.49	0.91	0.81	0.50	0.93	0.84	0.51	0.96	0.73	0.44	0.83
1	4.0	8.0	0.77	0.47	0.93	0.78	0.48	0.95	0.81	0.49	0.98	0.71	0.43	0.86
1	4.0	8.5	0.74	0.45	0.95	0.76	0.46	0.98	0.78	0.47	1.01	0.68	0.42	0.88
1	4.0	9.0	0.71	0.43	0.97	0.73	0.44	1.01	0.75	0.45	1.04	0.65	0.40	0.89
1	5.0	8.0	0.66	0.50	0.90	0.67	0.52	0.93	0.70	0.53	0.96	0.61	0.46	0.83
1	5.0	10.0	0.62	0.47	0.95	0.63	0.48	0.96	0.65	0.50	1.00	0.57	0.43	0.87

required for making one cubic yard of rammed concrete of variously proportioned mixtures with different coarse aggregates.

TABLE XII

Ingredients in One Cubic Yard of Concrete

(H. P. Gillette)

Sand voids 40%. Stone voids 45%. Barrel of Portland Cement containing 3.65 cu. ft. of paste. Barrel specified 3.8 cu. ft.

Proportions by Volume	1:2:4	1:2:5	1:2:6	1:2½:5	1:2½:6	1:3:4
Bbls. cement per cu. yd. concrete......	1.46	1.30	1.18	1.13	1.00	1.26
Cu. yds. sand " " " "	0.41	0.36	0.33	0.40	0.36	0.53
Cu. yds. stone " " " "	0.82	0.90	1.00	0.80	0.84	0.71
Proportions by Volume	**1:3:5**	**1:3:6**	**1:3:7**	**1:4:7**	**1:4:8**	**1:4:9**
Bbls. cement per cu. yd. concrete......	1.13	1.05	0.96	0.82	0.77	0.73
Cu. yds. sand " " " "	0.48	0.44	0.40	0.46	0.43	0.41
Cu. yds. stone " " " "	0.80	0.88	0.93	0.80	0.86	[illegible]
Proportions by Volume	**1:2:4**	**1:2:5**	**1:2:6**	**1:2½:6**	**1:2½:6**	**1:3:4**
Bbls. cement per cu. yd. concrete......	1.30	1.16	1.00	1.07	0.98	[illegible]
Cu. yds. sand " " " "	0.42	0.38	0.33	0.44	0.49	[illegible]
Cu. yds. stone " " " "	0.84	0.95	1.00	0.88	0.96	0.71
Proportions by Volume	**1:3:5**	**1:3:6**	**1:3:7**	**1:4:7**	**1:4:8**	**1:4:9**
Bbls. cement per cu. yd. concrete......	0.96	0.90	0.82	0.75	0.68	[illegible]
Cu. yds. sand " " " "	0.47	0.44	0.40	0.49	0.44	[illegible]
Cu. yds. stone " " " "	0.78	0.88	0.93	0.86	0.88	[illegible]

Measuring According to Voids. Where voids have been determined in both sand and coarse aggregate, the upper part of Table XII, which may be taken as representing average conditions, will be found useful.

Where the voids are the same as in the upper

part of Table XII, but the cement is measured loose in a box after dumping from a barrel, the quantities should be determined from the lower part of the table, as under such conditions a barrel of cement yields 4.4 cu. ft. of loose cement:

From the above table it will be seen that the following rule can be deduced: Add together the number of parts, and divide this sum into 10; the quotient will be approximately the number of barrels per cubic yard. Thus, for 1:2:5 concrete, the sum of the parts is 8, and $10 \div 8 = 1.25$ barrels of cement per yard, which is approximately equal to 1.30, the amount called for by the table.

THE OPERATION OF MIXING

The methods of mixing concrete vary greatly in detail, but may be broadly divided under two heads—**mixing by hand** and **mixing by machinery**. A further distinction is made among mixing processes, which may be either **batch mixing** or **continuous mixing**. Batch mixing, as the name implies, is an intermittent process, the ingredients being measured out, and the mixing done for only a "batch" or limited, definite quantity at one time. Continuous mixing is always done by machinery, the materials being fed to the machine in an uninterrupted stream, continuously during the entire mixing process. Batch mixing may be done either by hand or by machine. In machine mixing, both continuous and batch, the proportioning of ingredients is sometimes done automatically by the machine itself.

Hand Mixing vs. Machine Mixing. A choice as between hand mixing and machine mixing will depend chiefly on considerations of cost. Both methods will produce equally good concrete if sufficient care is taken to insure a thorough and uniform mixture. This is harder to make sure of with hand than with machine mixing, for, with the increasing fatigue of a day's work, and possibly that carelessness of workmanship which is bred of familiarity and monotony, a workman will naturally and inevitably show some falling-off in sharpness of attention and thoroughness of work toward the close of the day. Therefore, even with the closest supervision kept up by a good foreman, hand-mixed concrete is likely to be not quite so strong or so uniform in quality as the machine-mixed product —and this may be the case without any intentional neglect on the part of workmen.

Under any method of mixing—hand or machine—a careful and constant watch should be kept upon every detail of the process, to see that the proper relative quantities of the dry ingredients are being used, that the mixing is being thoroughly done, and that the concrete is of the proper consistency or degree of wetness. The same careful supervision should be kept up over every detail of the subsequent depositing, ramming, spading, drying, etc., of the concrete.

On small jobs—even those requiring as much as several hundred cubic yards of concrete—it is ordinarily much cheaper and more expedient to mix by hand. This is, of course, especially true where only a small crew—say two to four men—are available for the mixing, and where the work is often interrupted or frequent moving is necessary.

On large jobs it is more economical to mix by machinery, and this is the method generally adopted. The relative cost of the two methods usually depends on local circumstances, and these must be taken into consideration in each particular case.

Mixing Concrete by Hand

Details of the mixing process differ considerably among different workers, one preferring one method, another another. Usually the sand and cement are mixed first, dry, and afterward mixed with the stone before the water is added. Sometimes the cement and sand are first mixed into a mortar, before being mixed with the stone. The stone may be added to the dry cement and sand, or to the mortar, or vice versa. Sometimes the stone, sand, and cement are spread in successive layers, slightly mixed, and shoveled up into the form of a crater, into which water is poured, and the mass turned with shovels.

The essential thing in all cases is to see that the mass is turned a sufficient number of times to insure a thorough mixture.

The following account of the mixing of a small batch of concrete—condensed from a description given by Messrs. Wilson and Gaylord—will serve as a general illustration of details of procedure in ordinary practice on small jobs. The mixture is a 2-bag batch of 1:2:4 concrete, and two men are sufficient for the work.

The Concreting Plant. The batch will be mixed on a wooden platform known as a **concrete board.** For two men, this should be 9 feet by 10 feet for a 2-bag batch, or 12 by 10 feet for a 4-bag batch. Make it out of 1-inch boards, 10 feet long, surfaced on one side, using five 2 by 4-inch by 9-foot cleats to hold them together. The boards are so laid as to enable the shoveling to be done with, and not against, the cracks between the boards. The boards must be drawn up close in nailing, so that no cement grout will run through while mixing. If tongued-and-grooved boards are available, so much the better. Knot-holes may be closed by nailing a strip across them on the under side of the board. It is a good precaution against losing cement grout, to nail a 2 by 2-inch or 2 by 4-inch piece around the outer edge of the board. Often 2-inch planks are used in making concrete boards, but these are unnecessarily heavy, and very cumbersome to move (see Plate 2).

The concrete board is usually best placed as close as possible to the forms in which the concrete is to be deposited, but local conditions must govern this point. Choose a place giving plenty of room, near the storage piles of sand and stone (or gravel). Block up your concrete board level, so that the cement grout will not run off on one side, and so that the board will not sag in the middle under the weight of the concrete.

Make the runways for the wheelbarrows of good, strong plank 2 to 3 inches thick and 10 to 12 inches wide. They should be of liberal width —say at least 20 inches wide if lifted much above the ground—as this feature will help greatly to lighten and quicken the work.

In addition to board and runs as described above, the concreting outfit will include the following tools: **Shovels,** No. 3, square point; **wheelbarrows,** at least two being necessary for quick work (sheet-iron body preferred); **rake; water barrel;** water **buckets,** 2-gallon size; **tamper,** 4 by 4 inches by 2 feet 6 inches, with handles nailed to it, as shown in Plate 2; **garden spade,** or **spading tool** cut from a board and beveled to a thin edge at the bottom, as shown in the hands of the man on the concrete board, Plate 2, or in Fig. 8; and a sand **screen,** made by nailing a piece of ¼-inch-mesh wire screen 2½ ft. by 5 ft. in size to a frame made of 2-inch by 4-in. stuff.

Mixing. With the mixing board placed and the "runs" made, the concrete plant is ready.

First load your sand in wheelbarrows from the sand pile, wheel onto the board, and fill the sand-measuring box, which is placed about two feet from one of the 10-foot sides of the board. When the sand box is filled, lift it off, and spread the sand over the board in a layer 3 or 4 inches thick, as shown in Fig. 6 (a). Take the two bags of cement, and place the contents as evenly as possible over the sand. With the two men standing at points indicated, start mixing the sand and cement, each man turning over the half on his side of the line A-A. Starting at his feet and pushing the shovel away from him, each man takes a full shovel load, and turns the shovel over at the points marked 1 and 2. In turning the shovel, do not simply dump the sand and cement at the points marked 1 and 2 in the diagram, but shake the materials off the end and sides of the shovel, so that the sand and cement are mixed as they fall. This is a great assistance in mixing these materials. In this way the material is shoveled from one side of the board to the other, as shown in Fig. 6 (b and c). Fig. 6 (b) shows the first turning; and Fig. 6 (c), the second turning.

The sand and cement should now be well mixed and ready for the stone and water. After the last turning, spread the sand and cement out carefully; place the gravel or stone measuring box beside it as shown in Fig. 7 (a); and fill from the gravel pile. Lift off the box, and shovel the gravel on top of the sand and cement, spreading it as evenly as possible. With some experience, equally good results can be obtained by

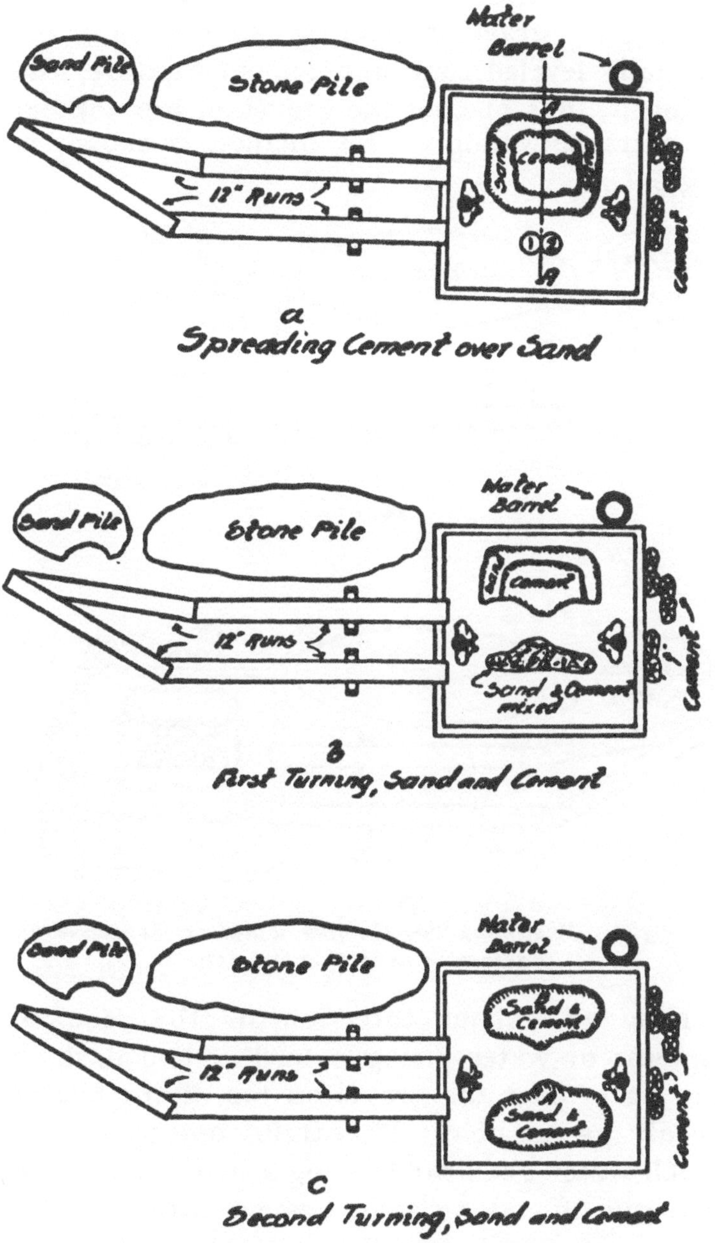

Fig. 6. Illustrating Hand-Mixing of Concrete.

placing the gravel measuring box on top of the carefully leveled sand and cement mixture, and filling it, thus placing the gravel on top without an extra shoveling. This method is shown in Fig. 7 (b).

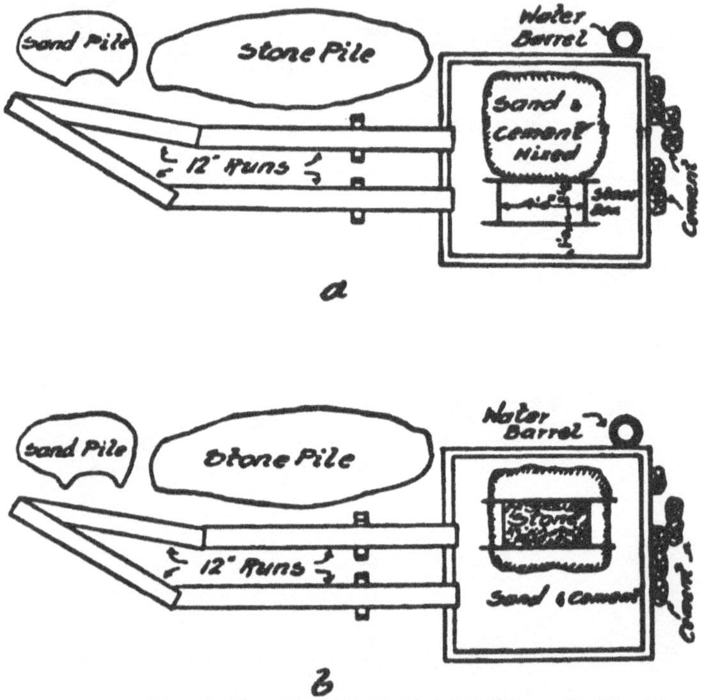

Fig. 7. Illustrating Two Methods of Adding the Coarse Aggregates in Mixing Concrete.

Now add about three-fourths the required amount of water, using a bucket, and dashing the water over the gravel on top of the pile as evenly as possible. Be careful not to let too much water get near the edges of the pile, as it will run off, taking some cement with it. This caution, however, does not apply to a properly constructed mixing board, as the cement and water cannot get away.

Starting the same as with the sand and cement, turn the materials over in much the same way, except that, instead of shaking the materials off the end of the shovel, the whole shovel-load is dumped as at points 1 or 2 (a, Fig. 6), and dragged back toward the mixer with the square point of the shovel. This mixes the gravel with the sand and cement, the wet gravel picking up the sand and cement as it rolls over when dragged back by the shovel. Add water to the dry spots as the mixing goes on, until all the required water has been used. Turn the mass back again, as was done with the sand and cement.

With experienced laborers, the concrete would be well mixed after three such turnings; but if it shows streaky or dry spots, it must be turned again. After the final turning, shovel into a compact pile. The concrete is now ready for placing.

When the sand and gravel are found already mixed, as in a natural bank, spread out the mixture of sand and gravel as much as the board will readily permit; add enough water to wet the gravel and sand thoroughly; spread the cement evenly in a thin layer over the sand and gravel; and turn over, as described previously, at least three times, adding the rest of the water necessary to get the required consistency while the materials are being turned. It requires some experience to work up a natural mixture of bank sand and gravel; and, if at all doubtful about the concrete made from it, first screen the sand from the gravel, and then mix in the regular way.

Number of Men. For the above operation, only two men are required, although more can be used to advantage. If three men are available, let two of them mix as described above, and let the third man supply the water, help mix the concrete by raking over the dry or unmixed spots as the two mixers turn the concrete, help load the wheelbarrows with sand and stone or gravel, etc.

If four men are available, it is best to increase the size of the batch mixed to a four-bag batch, doubling the quantities of all materials used. The cement board should also be increased to 10 feet by 12 feet. In this case, start the mixing in the middle of the board, each pair ot men mixing exactly as if for a two bag-batch, except that the concrete is shoveled into one big mass each time it is turned onto the center of the board.

When more than four men are available, the rest may place the concrete, make new runs, load wheelbarrows, etc., taking the concrete away from the board as fast as it is mixed. In this case, another small concrete board should be placed next to the big board, so that in the last turning the batch can be shoveled over onto the small board for placing, making room on the big board to mix the next batch. The small platform need be only just big enough to hold the pile of mixed concrete.

Measuring by Wheelbarrow. With a little practice, the sand and stone or gravel can be measured by the number of wheelbarrow loads almost as accurately as by means of the measuring boxes.

If the concrete has to be wheeled not over 50 feet, four experienced men should be able, on an average, to mix and wheel to the place of deposit about 10 four-bag batches of 1:3:6 concrete in 10 hours. And since, from Table VIII, one 4-bag batch will make 24 cubic feet of concrete, the four men should mix and wheel in this period of time about 240 cubic feet of concrete or $240 \div 27 = 8.8$ cubic yards. This estimate, however, is for the very simplest kind of concreting, and makes no allowance for the labor of supplying materials to the mixing platform or for building forms.

Mixing by Machinery. The details of machine mixing will be taken up later in connection with the description of the various types of concrete mixers, under the head of "Concreting Machinery and Tools."

To Regulate Setting and Hardening. As already explained, some regulation of the time of setting of cement can be accomplished by regulating the consistency of the mixture, a dry mixture setting more quickly than a wet one. Also, since concrete sets more rapidly in warm than cold weather, the setting may be regulated to some extent by simulating the proper weather conditions. This may be done by constructing a light framework over the job, covering it with canvas or other material, and regulating the temperature within by means of stoves. This method, however, is very expensive, and is used only in building construction, being employed then chiefly as a protection to the concrete during the setting process in freezing weather.

When it is desired to obtain a very rapid setting and hardening, as in making casts and ornamental forms, a 10 per cent solution of calcium chloride may be used instead of water, for mixing and pouring. This material is cheap, and can be obtained from any dealer in chemicals, or in large quantities from manufacturing chemists. A 10 per cent solution is obtained by dissolving 8 pounds of calcium chloride in 10 gallons of water. Artificial stone made in this way will harden, and can be removed from moulds or forms, in one-third the time required when only water is used in mixing. The chemical is said to have no injurious effect whatever on the strength or soundness of the cement.

DEPOSITING CONCRETE

Concrete should not be allowed to stand unused for any considerable length of time after being mixed, but should be placed or deposited at once. In no case should this be delayed beyond 20 to 30 minutes after the cement is first wet.

Effect of Remixing. To disturb the concrete after it has begun to set, will not only retard its setting, but may possibly prevent the development of its full, ultimate strength. This is particularly true where the cement is to be used in reinforced work, as remixing reduces the adhesiveness and binding power of the cement. If it is permissible to wait a long time—say several months—for the development of the full strength of the concrete, remixing of both Portland and natural cement mortar and concrete might be allowed, if done shortly after initial set has begun—say within an hour or two; but if, as is usually the case, it is desirable that the mixture shall acquire its full strength within a few days or weeks, then cement mortar or concrete which has begun to set before placing should be rejected.

If the mixing process is kept up for about two hours continuously before placing, it has been found that the strength of the mixture will be considerably increased, this being due, in all probability, to the more thorough blending and interpenetration of the constituents of the mortar or concrete.

Concrete is usually deposited in layers 4 to 12 inches thick. Unless otherwise specified, it should be placed in layers of about 6 inches.

Method of Placing. In the placing of concrete, the most important thing is to **handle it in such a way that its materials will not be separated.** The mass should retain its compactness and its even distribution of ingredients; the stone or gravel should never become separated from the mortar. Provided this precaution is taken, concrete may be handled and placed in any manner suited to the nature of the work.

If dumped into a trench or form from any considerable height, the work should be watched with special care to see that the heavier particles are not forced toward the bottom of the mass, and that the stone or gravel still remains in intimate contact with the mortar. Where depositing from a considerable height is necessary, a chute should invariably be used if the mixture is a "dry" one. A very wet mixture, however, is much easier to handle in such a case than a dry mixture, and the stone or gravel will not so readily be separated from the mortar.

Ramming or Tamping Concrete. In order to force out air-bubbles and obtain a thoroughly compact job, concrete should be rammed (or tamped) immediately after placing. The amount of ramming needed depends on the consistency of the mixture. If very wet, only the slightest tamping, if any, is necessary—merely enough to level off across the top. If much tamping is done to a very wet mixture, the particles of the coarse aggregate may be forced together in one part of the mass, and their uniform distribution be upset. Sometimes a very wet mixture, instead of being tamped, is simply **puddled** (stirred up), this being done with a piece of reinforcing bar or other convenient material.

A medium wet mixture should be tamped until thoroughly compacted, when it will quake like jelly.

A dry mixture must be very thoroughly tamped until the water is forced to the surface.

The tamper for medium or dry concrete may consist of a steel shoe or plate with a face about 6 inches square, to the back of which a handle of convenient length is fastened; or, for small work, a satisfactory home-made tamper may be constructed from a cylindrical block of wood into one end of which a handle is fastened. For work in very thin sections or between reinforcing steel members, a suitable rammer can be made from board 1 inch or so in thickness, the blade left at the lower end being about 8 inches high by 3 inches wide.

Spading. In order to insure a smooth, unbroken face on the concrete, it should be carefully **spaded** next to the forms, where the finished concrete will be exposed to view. And this is true whether the mixture be dry, medium, or very wet. The operation of spading consists in running a spade or flattened shovel down against the face of the form, and working up and down. This action causes the stone or gravel to be pushed back slightly from the form, and allows the cement grout to flow against the face of the form and fill any voids that may be there, thus making the face of the work present an even, homogeneous appearance. Where the narrowness of the concrete section, such as in a 6-inch silo wall, prevents the use of a spade, a 1-inch by 4-inch board, sharpened to chisel edge on the end, will do as well. Only sharpen on one side, and place the flat side against the form, as shown in Fig. 8. In the case of a dry mixture, spading must be done with greatest care by experienced hands, to get uniform results; but with a medium or very wet mixture, it is very easy to obtain first-class work; indeed, with a wet mixture, spading is required only as an added precaution against the possibility of voids in the face of the work, and is really necessary in few cases.

In spading, be careful not to pry with the spade, as this may spring the forms.

Depositing Concrete under Water. It occasionally happens that concrete must be deposited under water. In such cases, special precaution must be taken to prevent separation of the materials. The governing principle is to see that the concrete shall be disturbed as little as possible in being placed.

There are three methods used:

First, the concrete may be lowered under water to the place of deposit, in a closed bucket, which should be provided with an automatically

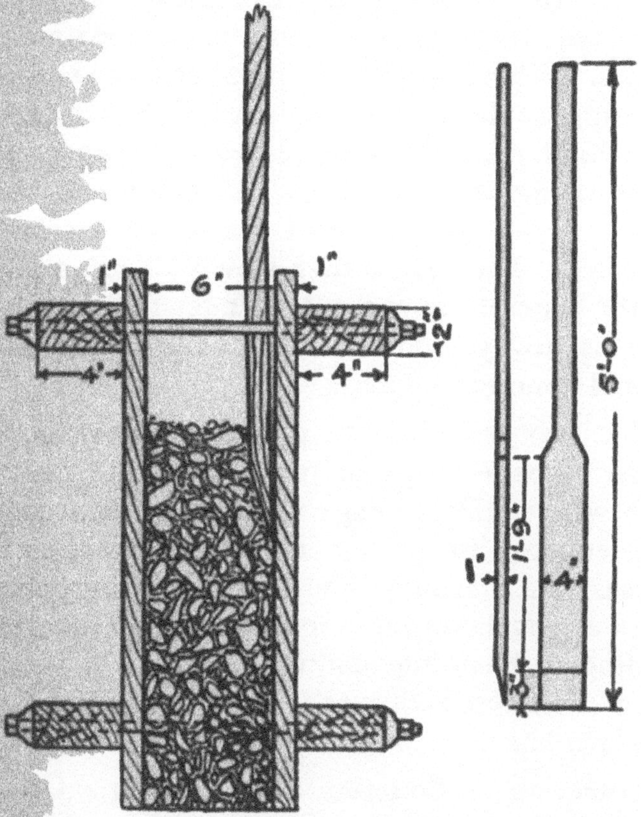

Fig. 8. Diagram Showing Method of Spading a 6-Inch Concrete Wall.

At right is a diagram giving details of wooden spading tool.

opening hinged bottom, allowing the concrete to slide out gently when its destination is reached.

Second, the concrete may be lowered to place in tubes. If only moderate quantities are to be deposited, a small tube—say 4 to 6 inches in diameter—with a removable cap at the lower end, may be used. This is first filled with the mixture, then lowered, the cap removed, and concrete fed through the tube in a continuous stream. Where large quantities are to be placed, requiring a tube of inconvenient weight for handling, this large tube may be lowered empty, the water rising therein. Into this chute the concrete may be dumped until the water has been displaced. The tube is then slowly raised, allowing the concrete to run out. Under this method, there is always more or less danger of badly washed concrete.

Third, concrete may be placed under water by means of paper bags or of cloth bags of open-mesh weave. Paper bags will retain their shape long enough to permit of proper placing in courses like masonry, and the paper soon loses all its strength through soaking, and allows the bonding or blending of the concrete. A similar union is effected through the open meshes of the cloth bags.

Protection of Concrete after Placing. Green concrete should not be exposed to the sun until after it has been allowed to set for five or six days. Each day during that period the concrete should be wet down by sprinkling water on it, both in the morning and afternoon. This is done so that the concrete on the outside will not dry out much faster than the concrete in the center of the mass; and it should be carried out carefully, especially during the hot summer months. Old canvas, sheeting, burlap, etc., placed so as to hang an inch or so away from the face of the concrete, will do very well as a protection. Wet this as well as the concrete. Often the concrete forms can be left in place a week or ten days; this protects the concrete during the setting-up period, and the above precautions are then unnecessary.

Cleaning the Concrete Board. When the day's work is done, carefully clean all the tools, especially the concrete board. Remove with a shovel all the loose cement, sand, and stone. Then scrub the board with a broom and water.

Bonding Old and New Concrete. Considerable care is necessary in order to secure a perfect bond or joint when new concrete is to be connected to old, or when a job of concreting interrupted for a considerable time is resumed, as when workmen quit work one day to take it up again the following day.

In ordinary work subject chiefly to compressive strains, as in foundations, it will be sufficient if the surface of the concrete laid one day is merely washed with clean water before laying on the new concrete the next day; but in walls and floors, or where tensile stresses will be developed, the surface to be joined should be thoroughly washed clean and soaked with water, and then painted either with neat cement (cement and water) or with a very thin, rich mortar consisting of 1 part cement to 1 part sand. If the old surface is very dirty, it may be necessary, after washing it as clean as possible, to pick it with a cold chisel in order to expose a perfectly clean and somewhat rough surface of concrete.

It will always help in the subsequent securing of a good bond with later work, if the old surface, after ramming or troweling, is scratched or roughened.

Where a water-tight wall is desired, as in a tank, if the concrete is laid in successive tiers and not by one continuous operation, special joints should be made which will give a mechanical as well as a chemical bond. They may be square or V-shaped in section, and either horizontal or vertical. Such joints are moulded by placing pieces of timber so that they will form grooves in the surface of the concrete when the last layer is placed each day, then removing the wood the following day, and coating the joint with neat cement or very thin, rich mortar before placing the next layer of concrete

Concreting in Frosty Weather

After concrete has once firmly set and hardened, frost has no effect upon it. It is different, however, if frost is allowed to attack the concrete while being mixed or placed, or during the setting process.

Many special tests and many investigations of

work in progress have been made, in order to determine the effect of freezing upon Portland cement concrete before it has had a chance to set. While some difference of opinion on this subject exists, it is the generally accepted conclusion of engineers that freezing does not materially affect the binding quality of good Portland cement, and that concrete laid in freezing weather will not be injured provided it does not freeze till after placing, and provided also that it is not subjected to any load until after it has been thawed out and allowed to set in the usual way. The freezing simply retards the process of hardening, which will again proceed under favorable conditions, until the concrete eventually develops its full strength.

On the other hand, alternate freezing and thawing before hardening is complete, is very apt to injure concrete. Accordingly, if concrete has once been frozen and thawed out before the action of hardening has commenced, it should be protected from freezing again until it has had a chance to harden sufficiently to withstand the action of subsequent frosts.

The freezing of a layer of concrete is very apt to prevent a good bond with another layer placed on top. In sidewalk and pavement work, freezing during construction is apt to cause a thin layer (about 1/12 inch in depth) to peel off, leaving the wearing surface rough; and a similar effect is sometimes seen on concrete walls. The injury, however, will not be deep-seated unless the whole mass has been subjected to repeated freezing and thawing before hardening is complete.

A green concrete mixture which can be easily frozen at a temperature below 32 degrees F. (the freezing point of water), should not be allowed to freeze, if this can be prevented. It is safest, therefore, to avoid mixing and placing concrete in freezing weather.

Protection from Frost. There are two general methods in use for protecting concrete from injury by frost, and expediting the progress of construction—namely, first, by using artificial heat; second, by using chemicals to lower the freezing point of the wet mixture so that the prevailing frost will not affect it. Sometimes both methods are employed on the same work.

Lowering the freezing point of the concrete is the simplest and cheapest, but probably not the best method of concreting in freezing weather. This method consists of adding some substance to the mixing water that will reduce its freezing point; but only those substances that have no effect on the strength and durability of the concrete can be used. Ordinary salt is most commonly used for this purpose; and experiments indicate that while the addition of a limited amount of salt retards the hardening somewhat and lowers the initial strength, the ultimate strength of the concrete is not affected by its use

Salt should be used only in plain concrete work, as its effect on reinforcing metal has not been established. Even when salt is used, it is important that the aggregates be free from lumps of frozen material, as it is impossible properly to mix such materials. Approximately one per cent by weight of salt to the weight of the water is required for each degree Fahrenheit below freezing; but more than ten per cent of salt—which is equivalent to about 13 pounds of salt to a barrel of cement—should not be considered safe, and this amount is not effective for temperatures lower than 22 degrees Fahrenheit.

The use of artificial heat may be accomplished in different ways—by inclosing the entire work under cover, and keeping it warm by heating the air within the enclosure; or by heating the aggregates or water, or both, before mixing. The former is the most expensive method, and is never employed except in building construction. The inclosing framework is lightly constructed of wood, covered with canvas or other material.

The best method of concreting in freezing weather is to heat the materials, and to protect the work until it has obtained sufficient strength to withstand the action of frost. Either the water, the sand and water, or the sand, stone, and water should be heated. The cement is usually not heated. Heating the materials accelerates the rate of hardening; lengthens the time before the material becomes cold enough to freeze;

and, in temperatures but little below freezing, will insure the hardening of the concrete before it can be damaged by frost.

For heavy mass work, thick walls, abutments, etc., it is not necessary to heat the stone except in unusually cold weather, but sand and water should be heated. If the forms are tight and made of heavy material, it will be necessary to protect only the top of the work; this may be done by covering with a canvas and running steam under it, or by covering with boards or paper and applying a covering of straw or manure. In no case, however, should fresh manure be allowed to come in contact with very green concrete, as it will discolor and spoil the surface. If the work, covered in the manner described, is protected from freezing for several days, it is sufficient, unless it has to be loaded immediately; but thin walls, light foundations, etc., should be protected on all sides in the manner pointed out above. For reinforced work it is necessary to heat all the materials but the cement; and the concrete should be hot when placed in the forms. Where the work must be placed in service as soon as possible, the only safe practice is to keep the surrounding temperature well above the freezing point until the work has thoroughly hardened.

The aggregates may be heated by being piled over sheet-iron pipes or drums in which fires are built, or over steam pipes laid through the bottom of the storage bins. The water may be heated by exhaust steam or in any other convenient way. When a mechanical mixer is used, it also is sometimes kept warm, by means of steam coils outside and jets inside.

Concrete increases in strength but very slowly in cold weather; and for this reason, forms should be left on as long as possible, and care taken not to load a structure too soon. Just how old the work should be before removing the forms and subjecting it to its load, cannot be stated, as this will depend entirely upon how fast the concrete hardens. Careful inspection of the structure is necessary before removing the forms and applying the load; and it must be remembered that frozen concrete closely resembles thoroughly hardened concrete in appearance, and, when broken, frequently shows a fracture through the aggregate, although upon thawing it may have but little strength.

If the aggregate is porous, it should be well soaked in water before coming in contact with the cement; otherwise it will absorb the water necessary for mixing and for developing the full activity and strength of the cement.

Efflorescence is the name commonly applied to the whitish scum sometimes appearing on stone, brick, terra-cotta, and concrete work, due to the leeching out of lime or other soluble chemical salts. If the scum is due to excess of cement it is known as **laitance.** The Sylvester process, consisting of alternate applications of hot soap and alum solutions, has proved effective in its removal. A wash of dilute hydrochloric acid (one part acid to forty parts water) will also be found effective. Efflorescence can be prevented by waterproofing the exterior surface of the wall after the concrete has hardened.

Crazing or Hair Cracks. One of the troubles which for a time mystified concrete workers, was the appearance of fine cracks—known as crazing or hair cracks on concrete surfaces, frequently long after the concrete had set hard. Investigation, however, showed that this disfigurement was entirely superficial. Hair cracks are confined to the surface, being generally of no greater depth or width than a coarse hair; they are no indication of weakness in the concrete. It has long been known that very wet concrete is more apt to craze than dry concrete is. Dry concrete is objectionable to some for the reason that it is lacking in strength and density; and on that account its use is not advocated even though hair cracks are to some extent avoided. Experiments have demonstrated the fact that in wet concrete a portion of the very finest particles of cement is carried to the surface by the action of the excess water which is being absorbed by the atmosphere. This excess water is to a large extent drawn from the interior of the concrete to the exterior, carrying with it the finer particles, which, being deposited on the surface, form a richer mortar than is contained in the interior. Under certain conditions these finer particles on the surface practically form a coating of neat cement. Neat cement and rich mortars are found to be much more liable to crazing than mortars containing a larger proportion of sand or finely crushed stone. The cracks are due entirely to a contraction of the surface, and not to any contraction of the interior.

The trouble may be almost entirely avoided by seasoning the concrete in a very moist atmosphere or under a cover of very wet sand, or, in the case of blocks, by immersing entirely for a considerable time in water.

In the past, a practice that has been partially effective in overcoming the trouble has been to brush off the surface with a stiff steel brush, or to scrub it with a cement brick and wet sand or carborundum stone, thus partially removing the neat cement face in which the cracks develop. This remedy, however, is only temporary, the cracks being likely to appear months later. The brushing or scrubbing is merely an assistance, the real remedy being to keep the surface continuously and thoroughly wet as long as possible. For permanent results, the only effective remedy is to remove from the surface, as soon as the forms or moulds are taken down, as much of the excess of neat cement as possible; and then to allow the concrete to harden under conditions practically the same as when hardened under water—keeping it constantly wet during the hardening process by sprinkling, by covering with very wet sand, by inclosing in a very moist atmosphere, by immersing under water, or by some other convenient method of a similar character.

STRENGTH OF CONCRETE

Ice only an inch thick covering a stream or lake will carry safely the weight of a loaded team, whereas a covering of snow a foot thick will not support the weight of a child. What makes the difference in strength between the ice and the snow? Both are of the same chemical composition. The only difference between them—and the only cause of weakness in the snow as compared with the ice—is the presence, in the snow, of a large proportion of air-spaces or voids, greatly lowering its density.

It is much the same with cement mortar and concrete. The **quality of the materials** entering into the mixture will, of course, affect its strength. Trap rock or granite, for example, will give a stronger concrete than sandstone; and a good quality of cement will prove superior to a poorly ground and poorly calcined product. But, other things being equal, the most important factor affecting the strength of concrete is its **density**, and this, in turn, depends upon the **quantity of cement used**—that is, the amount measured proportionately to the volume of the mixed concrete; upon the sizes of the aggregates and their proportional grading of coarse and fine; and upon the thoroughness with which the mixing is done and the ingredients compacted together.

With given proportions of cement, sand, and gravel or crushed stone, the strength will depend chiefly on the quality of the materials and the thoroughness of the mixing. The loss of strength due to the use of improper sand may amount to as much as 50 per cent; and a 25 per cent loss of strength may result from carelessness or laxity of methods in mixing.

The strength of concrete as shown under direct compression affords a guide to the safe vertical loads that may be placed upon it. The results of tests by Taylor and Thompson are given in Table XIII, showing the safe vertical loads which may be placed upon Portland cement concrete of various mixtures after one month's setting, where the height of the column or mass is not over, say, twelve times its least horizontal dimension.

TABLE XIII

Safe Loads for Portland Cement Concrete under Direct Compression

Proportions	Pounds per Sq. In.	Tons per Sq. Ft.
1:2 :4	410	29
1:2½:5	360	26
1:3 :6	325	23
1:4 :8	260	18

In case of large mass foundations, values one-eight greater than those given in the above table may be taken.

Where the concrete mass is subjected to vibrating or pounding loads, the values given in the table should be reduced one-half.

The **tensile strength** of concrete is very much less than the **compressive strength**, ranging from one-tenth to one-fifth of the latter. It is also a more uncertain quantity. The tensile strength of good concrete suitable for reinforced construction, well aged, ranges from 200 to 500 pounds per square inch.

Weight of Concrete. The weight of a cubic foot of concrete is ordinarily considered as 150 pounds, its variations above or below this depending on the specific gravity of the materials used and the compactness of the mixture. Cinder concrete weighs only about 110 pounds per cubic foot, and its strength varies from one-half to two-thirds that of stone or gravel concrete.

Simplest form of seat—Edges should be moulded or rounded, and supports should have a moulded base.

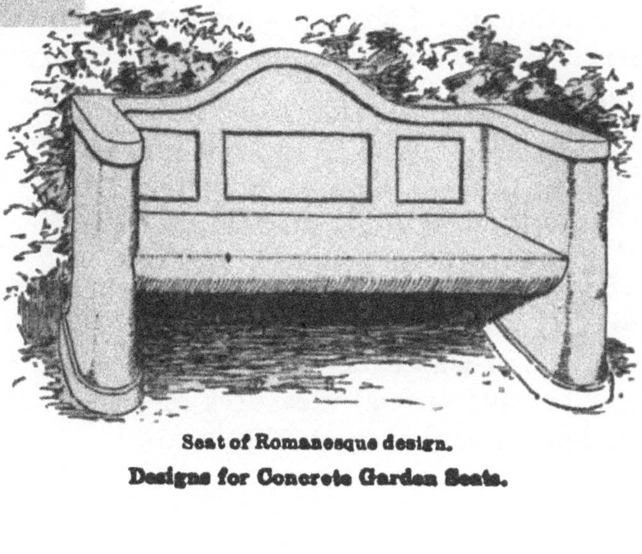

Seat of Romanesque design.

Designs for Concrete Garden Seats.

Waterproofing of Concrete

The development of a satisfactory method of waterproofing concrete is one of the most difficult and complicated problems that have exercised the minds of engineers since this material first came to be extensively used as a factor in modern construction. In its entirety, the problem presents a great multiplicity of conditions to be met and ends to be achieved. In some instances—as, for example, in the construction of reservoirs, tanks, and irrigation works—it is necessary to render the structure impermeable to water, even under abnormal conditions of pressure. In other cases—as in the erection of foundations and walls for buildings, where only normal conditions of ground-pressure and atmospheric moisture are to be met—it is necessary not merely to prevent seepage of water through the structure, but to avoid all tendency even to the absorption and retention of moisture from the air, with their resultant unhealthful dampness. And in still other cases—as in some concrete buildings both of monolithic and of block construction, where all danger of the penetration or absorption of water has been obviated—it is necessary to take special measures to prevent condensation of moisture on the inside.

In view, therefore, of the great complexity of conditions that characterize the problem, it is highly improbable that any single method of waterproofing can ever be said to meet in the most effective and at the same time the most economical way all possible conditions, and to be unqualifiedly "the best" for all possible cases. As with other engineering problems, the partieular features of each individual case call for individual attention and treatment.

Concrete is not the only porous building material that has been known to mankind; there is abundant authentic evidence to show that it has no monopoly of the unenviable quality of absorbing moisture. Concrete houses are not the only ones haunted with the specters of reeking dampness and mould and other un-hygienic ghosts of bygone times; nor are they the most difficult to flood and flush and brighten with the health-giving stream of modern sunshine and fresh air.

Wood, brick, stone, and, in fact, other building materials in general, with the notable exceptions of steel and glass, share in varying degrees with concrete the quality of being decidedly porous and absorbent; and it is altogether probable that if the principles of grading and proportioning to give mixtures of maximum density had been fully known and appreciated in the early days of the concrete building industry, the cry of "dampness" as regards concrete houses would never have been raised. A concrete house, properly built—and it is possible to build it so—is as dry, as light and cheery, as easily regulated in temperature, as well ventilated, and as healthful, as one built of any other material.

The importance of waterproofing is emphasized by several considerations. In the case of structures designed to retain water, it is, of course, self-evident. In the case of buildings designed for habitation or for commercial or industrial use, the prevention of abnormal dampness is essential to the health and comfort of the occupants, and has an important bearing on the life of the structure itself. Where the natural conditions of porosity and absorption common to structural materials in general are allowed to prevail unchecked, water is drawn into the foundations from the surrounding soil, and absorbed by the walls from the atmosphere. By capillary attraction, it spreads so as finally to permeate the entire structure, actively attacking and in time corroding and destroying the more or less unstable materials of which the structure is built, and producing damp, clammy walls which constantly foster and disseminate disease. The adoption of an efficient method of damp-proofing and waterproofing is therefore of vital importance, not only preventing the gradual decay and disintegration of structural materials, but going further to establish better hygienic conditions for the benefit of all classes. A practical solution of the problem, moreover, giving effective but reasonably cheap methods of waterproofing adaptable to varying conditions, will greatly promote the cement industry by opening up an enlarged field of usefulness for concrete, allowing it to be used in a great variety of work to which it formative characteristics peculiarly adapt it, but for which, without protection, it would be of little value.

METHODS OF WATERPROOFING

The various methods devised for the waterproofing of concrete may be classified under two broad divisions as follows:

1. **Mass or Integral methods**, in which the mass of the concrete itself is waterproofed prior to setting.

2. **Methods of external or surface treatment**, in which a protective, impervious covering is applied to the surface of concrete or placed between successive strata during the erection of the structure.

Both of these methods may be—and in many instances are—effectively combined in the same construction.

Each of the above classes into which waterproofing methods are divided may be further subdivided into a number of special methods differing from one another in marked characteristics.

Integral methods, for example, comprise three great subdivisions as follows:

(a) The use of carefully graded and proportioned and thoroughly compacted mixtures to give maximum density to the concrete.

(b) The introduction of special waterproofing materials into the mixture; these may be in the form of a powder, a paste, or a liquid.

(c) The special treatment of Portland cement itself during the process of manufacture, so that as it comes from the mills it already contains all the waterproofing elements required, with appropriate aggregates, to give an impervious mixture.

Surface-treatment methods of waterproofing may be subdivided according to the materials used:

(a) One method that has been very widely adopted and has been found of great practical efficiency, even under conditions of great severity, consists in the use of asphalt or of asphalt- or tar-product mixtures, either used alone or in combination with felt or similar waterproofing fabrics. This method will be more fully described later.

(b) A method that has been proved effective where the requirements are not very severe consists in plastering the wall—immediately after the removal of the forms, if possible—with a coat of very rich cement and sand mortar. This coating is usually from one-half to three-quarters of an inch thick, and lime paste is sometimes added for smoothness in working. It is essential that the coating be well troweled and smoothed off, preferably with a wooden float, as the hard skin developed by such treatment is practically waterproof. A neat cement wash is sometimes used, applied with a brush. On horizontal or inclined surfaces, the surface of the concrete should be worked and troweled, giving the effect of a cement coating.

(c) Another method that has given satisfactory results in preventing water penetration and absorption under conditions of moderate pressure, consists in the application of washes that depend, for their effect, upon chemical action. Perhaps the most widely used example of this method is what is known as the Sylvester process of waterproofing, which consists in the alternate application of hot solutions of castile soap and alum. The application is preferably made while the concrete is still "green;" but the process may be used for the waterproofing of old walls either of cement or lime plaster, stone, brick, or other more or less porous material. The walls in every case should be clean and as dry as possible, and the temperature of the air not lower than 50 degrees F. The soap is first dissolved in water—three quarters of a pound of soap to a gallon of water—and, while boiling hot, is spread over the surface with a flat brush. Twenty-four hours later, the alum solution—one pound of alum to eight gallons of water—should be applied in the same manner, at a temperature of about 65 degrees F. The process is repeated every twenty-four hours, four coats being usually sufficient to effect the desired result. From the chemical combination of the soap and alum, an insoluble compound is formed which effectively fills the pores of the structure to an appreciable depth and checks the penetration of moisture.

A process similar to the above in its action has been used for making mortar non-absorbent. Powdered alum (one per cent by weight) is thoroughly mixed with the dry cement and sand; and about one per cent of any potash soap (such as ordinary soft soap) is dissolved in the water used in mixing the mortar. A wash consisting of five pounds of alum dissolved in two gallons of water, to which one pound of concentrated lye has been added, has been successfully used on green concrete surfaces. Boiled linseed oil, applied in successive coats until it ceases to be absorbed, has also been used effectively.

(d) In addition to the above methods of surface treatment, there are on the market a large number of special patented waterproofing compounds and processes of more or less merit, manufactured according to formulae, and sold under various proprietary trade names.

It must be understood that there is no hard and fast line of demarcation between the integral and surface methods of treatment. A method that is in its nature essentially integral, involving the incorporation of special materials throughout the mass of that portion of the concrete which is directly affected, may, in the manner of its application, have all the appearance of a purely surface method of treatment, the waterproofed material being spread upon the sur-

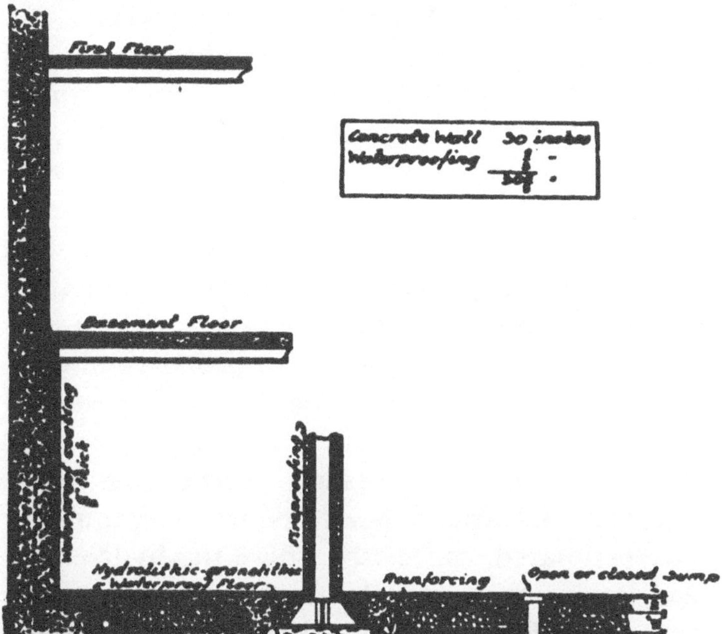

Fig. 9. One Method of Waterproofing a Building.
Walls and floor are placed monolithic; waterproof coating applied on inner surface.

face or deposited between successive strata in much the same way as asphalt and other bituminous products are applied.

There are some points of vital bearing on the problem of waterproofing, regarding which there is no dispute. It is universally conceded, for example, that proper proportioning of ingredients, proper grading of aggregates, thoroughness of mixing, and careful methods of placing concrete are factors of the utmost importance. **The densest concrete that can be made with the given materials will be the most impervious to water.** Also, the richer the mix, where no special compounds are used, the more nearly waterproof will the concrete be. It is evident, therefore, for practical reasons, that, for waterproofing work where the concrete alone is depended upon, a tendency to excess of fine materials is better than the reverse. Especially in monolithic construetion, it is now generally conceded that a wet mixture, a rich concrete, and a proportioning of aggregates to give great density, are essential for securing a waterproof structure.

The amount of pressure exerted by water against walls and beneath floors is frequently underestimated, especially where the hydrostatic head is low or is prevalent for but a short time after a rainfall.

INTEGRAL METHODS OF WATERPROOFING

Under this head, as before explained, come all those details that relate to the securing of maximum density in the concrete itself by the use of carefully graded and proportioned and well-compacted mixtures. In addition, however, this heading embraces a great variety of special products sold under various proprietary trade names, and intended for direct incorporation in the concrete mass. The majority of these on the market are in powder form, but some are in the form of a solution.

SURFACE TREATMENT METHODS OF WATERPROOFING

Under this head we find the widely used processes of asphalt and tar-product waterproofing, together with a long list of special proprietary

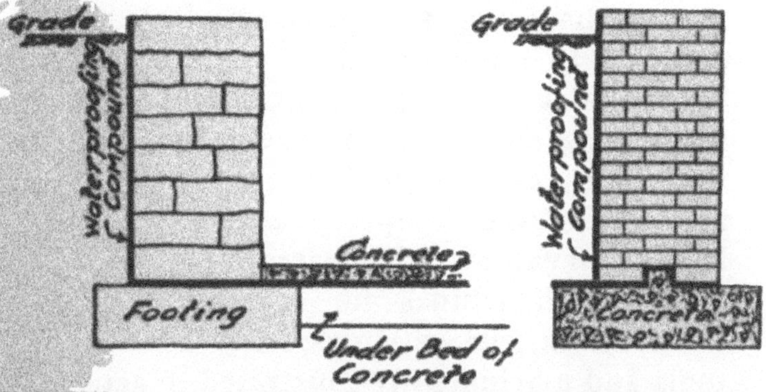

Fig. 10. One method of Waterproofing Stone and Brick Buildings. Waterproofing applied outside walls and over footings.

compounds (many of which are asphaltic or bituminous in their nature) sold in paste or liquid form for direct application to the surface; also various special brands of weather-proofing and damp-proofing paints.

Asphalt Waterproofing. Asphaltic coatings, either alone or in combination with felt or other fabric, may be applied to the surface or between two sections or layers of the concrete structure.

poured hot between the face of the main a thin protecting wall of brick or conc in front of it and carried up as the ·w gresses.

TOOLS AND MACHINERY FOR CRETE WORK

With the skill and ingenuity charact American enterprise in general, desig inventors have developed a great v special tools, machines, and devices fo plishing the different operations necess mixing, handling, and working of conc addition to numerous hand-operated t list includes crushers and grinders; scr separators; mixers; automatic measurin for water, cement, and aggregates; a and pneumatic tampers; engines fo hoists; elevators; wheelbarrows; dum and cars; machines for making blo posts, brick, tile, shingles, etc. Some tools and machines are illustrated in ac ing figures and plates.

The selection of the equipment neces depend on considerations of economy ciency; and these, in turn, will be fou pend on the size of the job, on the qu

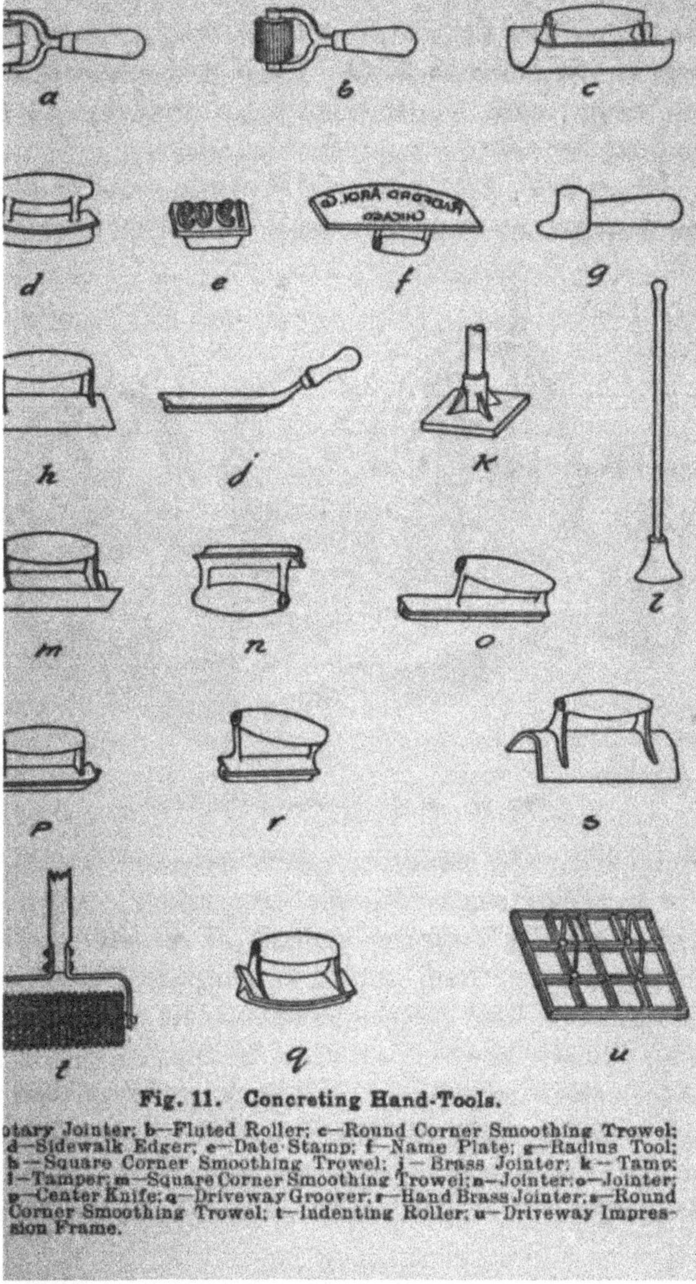

Fig. 11. Concreting Hand-Tools.

otary Jointer; b—Fluted Roller; c—Round Corner Smoothing Trowel; d—Sidewalk Edger; e—Date Stamp; f—Name Plate; g—Radius Tool; h—Square Corner Smoothing Trowel; j—Brass Jointer; k—Tamp; l—Tamper; m—Square Corner Smoothing Trowel; n—Jointer; o—Jointer; p—Center Knife; q—Driveway Groover; r—Hand Brass Jointer; s—Round Corner Smoothing Trowel; t—Indenting Roller; u—Driveway Impression Frame.

be mixed per day, and on local conditions peculiar to the work in hand. Each job, particularly on large work, constitutes a problem by itself, calling for careful analysis of details.

In general, it may be said that the plant should be located as near as possible to the place of

Fig. 12. Single Friction Drum Hoist.

depositing the concrete; uninterrupted facilities for handling materials, charging mixers, and distributing the concrete should be provided; the cycle of operations should be simplified as much as possible, and systematically maintained; machine parts should in all cases be capable of ready duplication to prevent delays from breakdowns or wearing out of parts; a plant should be adapted not only to the original job for which it is selected, but to subsequent work of other kinds; and, on small jobs, to depend on hand labor is usually more economical than to invest large sums in machinery.

Concrete Mixers. Concrete mixers comprise two general classes—continuous mixers and batch mixers. With the "continuous" type, the feeding of the materials, the mixing, and the discharge of the concrete form one continuous process which may be carried on indefinitely. With "batch" mixers, on the other hand, only certain measured quantities of cement, sand, etc., are fed to the machine, and only a definite quantity or "batch" of concrete is made and discharged at a single operation, the process being repeated as often as may be necessary. Some mixers are designed so that they can be adjusted to do work either continuously or in batches, and thus combine features of both classes of machines.

In continuous mixers, the general mechanical principle is that of a long screw or pug-mill, the mixing being done in a long drum by means of blades or paddles mounted on revolving shafts, which, somewhat like the thread of a screw or the blades of a propeller, agitate the materials and at the same time shove them along toward the discharge end of the drum. In the combined type, the discharge opening in the drum may be closed, the feed cut off, and only a batch mixed at a time.

Concrete mixers are further distinguished according to their mechanical construction and operation, into three classes—namely, **gravity** mixers, **revolving** or **rotary** mixers, and **paddle** mixers.

In the gravity type, as the name indicates, the operating force is that of gravity. The materials are mixed by simply being thrown down an inclined chute. On the interior surface of this chute are fastened projecting members in the shape of pins or blades, which are successively struck by the materials in their descent. These mixers are usually constructed of sheet steel, and may be had in sectional form adjustable to the greater or less height of the raised platform from which the materials are dumped. Mixers of the gravity type were the earliest mechanical mixers used.

Rotary mixers comprise a great variety of commercial types. In general, they consist of a hollow, revolving drum or box usually power-driven, but sometimes operated by hand. In some cases the shape of the box or the form of the plates composing the drum, combined with the manner of mounting, is depended upon to do the mixing as the machine revolves, without the assistance of any interior paddles or blades. In other cases, blades or reflectors are mounted in the interior of the drum in positions calculated to intercept the material, cut through it, and throw it about from side to side, thus mixing it thoroughly. Many machines of this type can be filled and discharged while running.

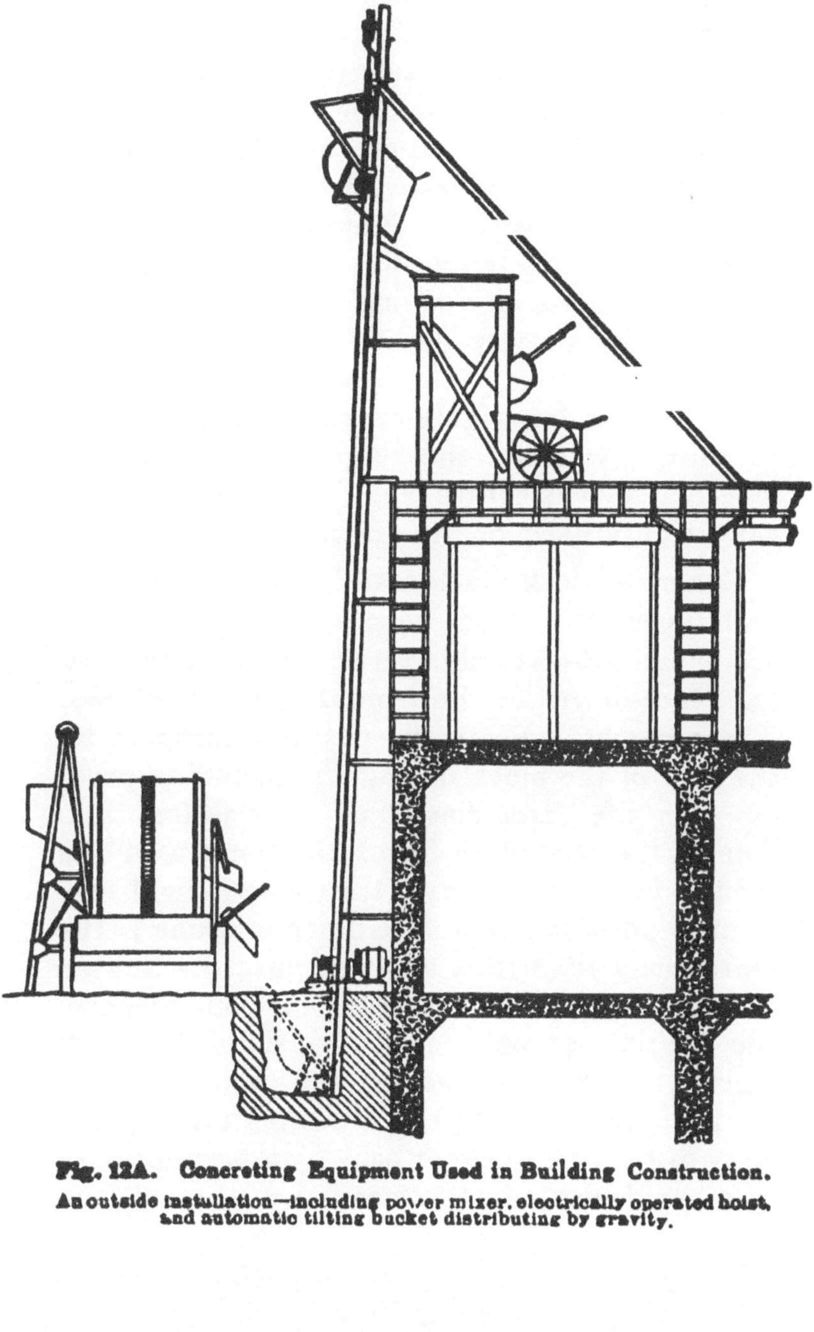

Fig. 12A. Concreting Equipment Used in Building Construction.

An outside installation—including power mixer, electrically operated hoist, and automatic tilting bucket distributing by gravity.

Fig. 12A. is a diagram illustrating a concreting equipment operating outside the structure; other installations of this same type are adapted to inside work. The installation shown in the figure includes a power mixer with elevator for automatically hoisting and dumping the charging bucket; also an electrically operated extension hoist, with automatic tilting skip or bucket discharging by gravity on the different floors as the structure rises. The concrete is discharged into a bin having a gate operated by hand-lever as shown, whereby the supply to the carts or wheelbarrows is controlled. The skip automatically rights itself on its return.

Concrete Block Machines. There are two general types into which concrete block machines are classified—namely, vertical-faced machines and face-down or horizontal-faced machines. These names indicate the position occupied by the face of the block during the moulding process. In the "face-down" or "down-face" machines, the face-plate forms the bottom of the mould; in the other type, this plate is held in a vertical position. The face-down machine is the more easily adapted to applying specially treated faces to blocks. If a vertical machine is used, the special facing (consisting simply of rich mortar, which may be colored, waterproofed, etc.) is placed while the mixture forming the body of the block is kept back by a parting plate or other device. In the face-down machine, the facing can be placed directly in the bottom of the mould, and the coarser body mixture be placed above it.

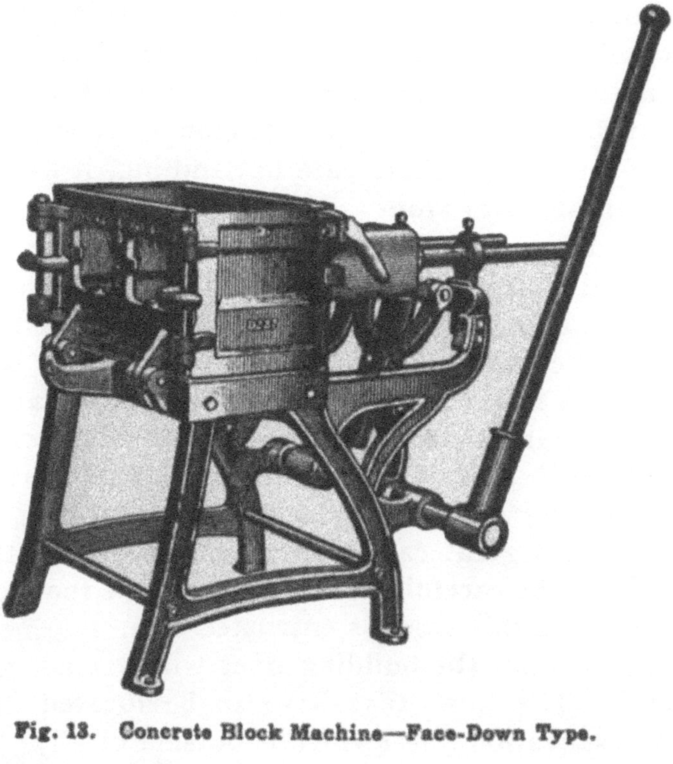

Fig. 13. Concrete Block Machine—Face-Down Type.

Cement Brick and Drain-Tile Machines. Many of the companies manufacturing concrete block machines have also developed special machines for the making of cement brick, drain-tile, Y-sewer pipe, T-sewer pipe, and drain-tile connections; also for making cement shingles.

CONCRETE FOUNDATION WORK

Except in a few localities where native stone is to be had very cheaply, all foundation walls for residences as well as for larger buildings are coming to be built of concrete. Builders have found that for strength, warmth, and enduring qualities, foundation and basement work in this material is far superior to brick or to wood piles; and for economy and ease in handling, it has an advantage over stone.

Building Lines. After the building site has been determined exactly—either in accordance with the architect's drawings or determined by the soil conditions, elevation, grade, etc.—it is in order to stake it out. This is done by placing stakes outside of each corner and connecting them with cords to guide the workmen in their excavating. It is very important that this be done with great care. Even in small buildings, it should be carefully attended to; while for large structures this work is entrusted to an engineer, who lays out the building lines with transit and level. The lines that have to be located are: the excavation line (which is outside of all); the face line of the basement wall; and, for masonry construction, the ashlar line, which indicates the outside face of the brick or stone wall.

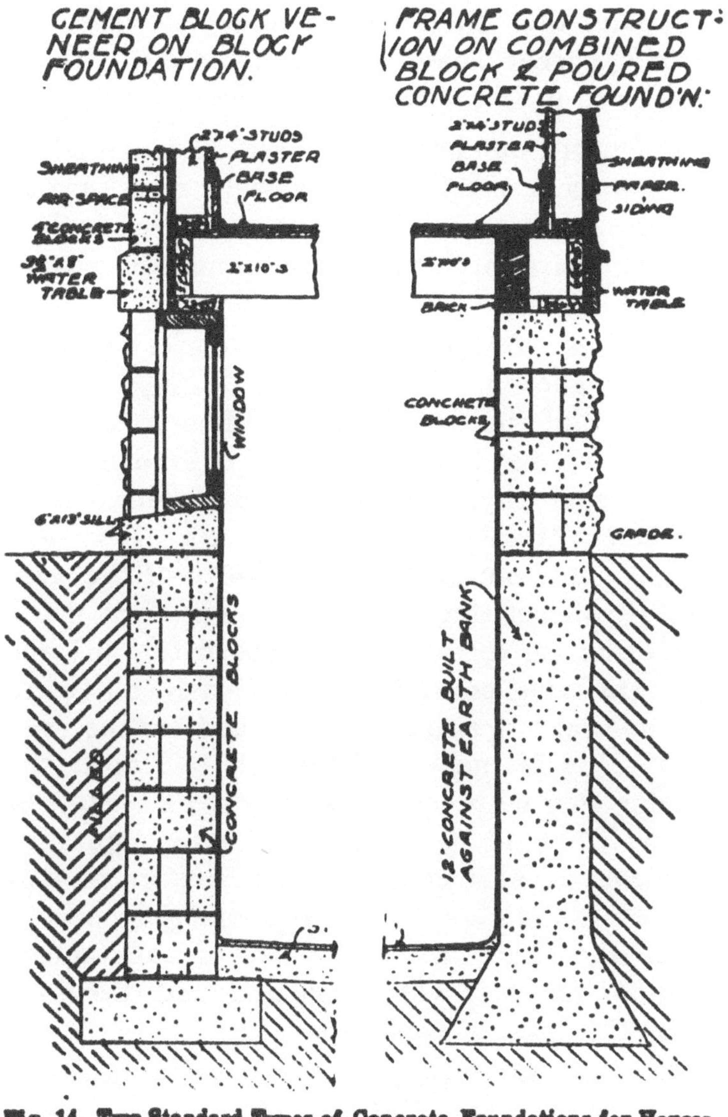

Fig. 14. Two Standard Types of Concrete Foundations for Houses.

The main rectangle of the plan is laid out first; and then the supplemental rectangles—as for ells, porches, bays, etc.—are laid out with reference to it, in their proper places.

Types of Foundations. There are a number of types of concrete foundation walls now in general use for houses. Two are illustrated in Fig. 14. They are: first, the entire foundation wall of cement blocks; second, the combination wall (poured concrete to grade, and blocks or dressed stone above).

A wall of the first kind is shown to the left. Excavation for foundation of this kind is made in the usual way, deep enough to provide a footing below frost (3 to 5 feet down). It is well to make the footing twice the width of the wall, and 10 inches thick. If the soil is firm, as it should be, no forms will be needed for this, the concrete being poured into the trench to harden.

A special large-size block is good for the wall, 8 by 12 by 24 inches. These are laid up in the regular way with cement mortar. The inside of the wall should be finished off with a quarter-inch coat of neat cement. It is important that cement foundation walls be thoroughly waterproofed. There are a number of satisfactory methods for this, which will be found discussed in another part of this book.

The second type, or combination wall, is shown to the right in Fig. 14. This is very good, especially where the soil is firm; for, in that

Excavation is carefully made, stopping just at **the outside** foundation line; the bank is hol**lowed back** in under, for a sloping footing below **frost; and** the inside forms are set up. Con**crete,** composed of 1 part cement, 2½ parts **sand, and** 5 parts crushed stone or gravel, very **thoroughly** mixed, is then carefully shoveled in **and tamped** solid. This wall, if desired, may be **waterproofed** by one of the special methods al**ready** described under "Waterproofing of Concrete." When this foundation has hardened sufficiently, the upper wall of blocks or dressed **stone is** laid up in the regular way.

Fig. 14 shows also two methods of framing for the superstructure—one for an ordinary **frame** building, standard construction, and the **other for** a frame building veneered with four-**inch-thick** concrete blocks. These should be **secured** to the framework, either with patent an**chors** or with large spikes driven into the wood **with the** heads built into the joints.

An economical foundation **wall** sometimes **used** where the building code prescribes thick **walls,** is a combination of hollow block and mo**nolithic** construction. Its economical features **are not** confined alone to the saving of concrete, **but include** the forms also, as scarcely any form **is necessary** for the footing. After the footing **has hardened,** the piers are built, and then the **curtain** walls, requiring but a few forms, which can be used over and over again without resawing or wasting lumber. Fig. 15 shows the arrangement. Piers 6 or 8 feet apart are erected, using a grooved block. Between these piers the curtain walls are placed after the piers become hard.

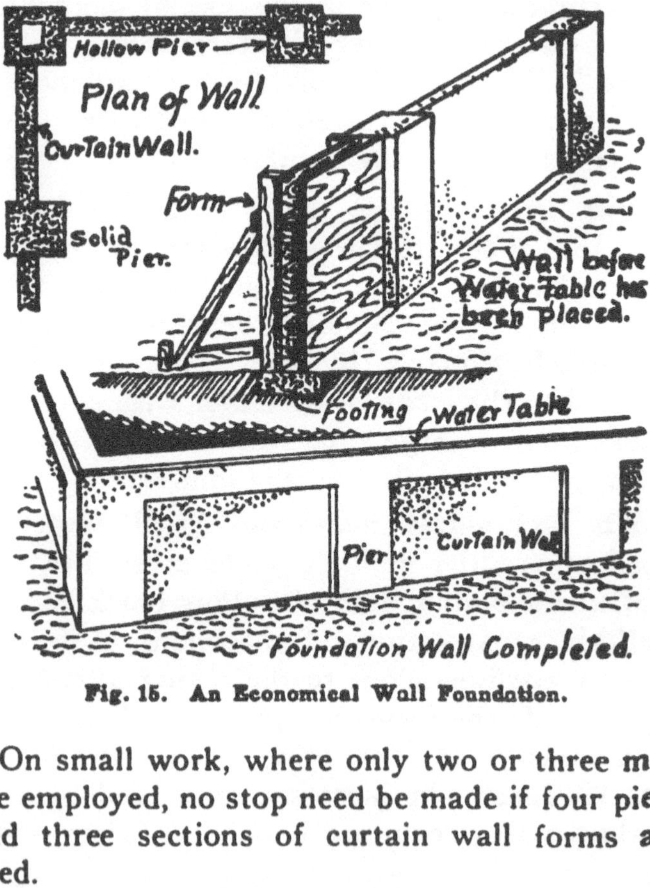

Fig. 15. An Economical Wall Foundation.

On small work, where only two or three men are employed, no stop need be made if four piers and three sections of curtain wall forms are used.

The water-table is made after the piers and curtain walls are self-sustaining.

By the use of hollow blocks for piers and monolithic curtain walls, this method of construction is surprisingly rapid and effects a great saving of cost, especially in localities where the hauling adds much to the cost of concrete.

The appearance of this wall is preferable to that of the straight, plain type. Besides, when building codes class concrete with rubble stone walls in thickness, only the piers need be the thickness required, while the curtain wall is usually acceptable if six inches thick. Walls of this type have been made as light as four inches, and have stood every test.

With this method, a single wagon carries all forms and tools from one job to another; the cost of the forms, made of surface lumber, is about $18.00, while the waste of lumber on a complete form for a dwelling foundation wall 30 by 40 by 7 feet high, for a 12 or 18-inch wall, will be $25, to say nothing of discoloring about $150 worth of good lumber.

By adding about one pound of ultramarine blue to each barrel of cement used for curtain walls, a beautiful effect is obtained, as it gives the piers and water-table a lighter color and more massive appearance.

Waterproof Cellar near a Stream. It is frequently necessary to build a cement cellar close to a stream where water is liable to seep in from the bed to the bottom of the cellar when the water rises during the spring freshet. Waterproofing for basements so located must be strong and well made, as it must **resist pressure.** A good method is to apply the waterproofing on the outside of the wall, covering same with cement plaster as shown in Fig. 16. On the bottom, apply the waterproofing on the concrete body, and cover wth the half-inch cement finish (wearing coat). Care must be taken to cover

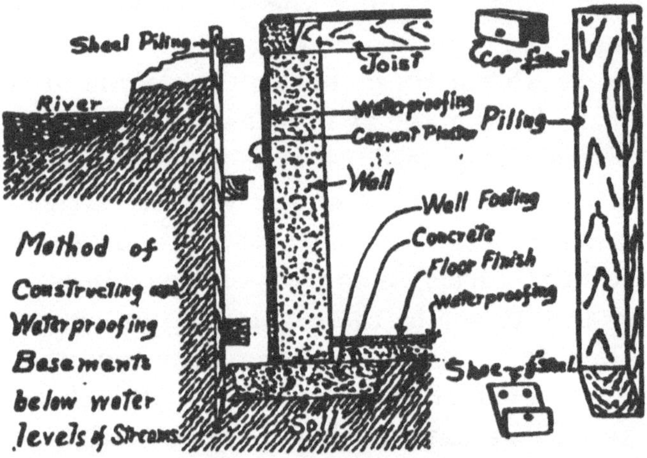

Fig. 16. Waterproofing Cellars below Water Level.

the entire surface to make it absolutely water-tight. On old walls and floors, the waterproofing must be applied on the interior.

The use of sheet piling made of wood, with a sheet-steel shoe for each plank, and a driving cap of sheet steel as shown, saves labor, time, and cost on even the smallest job requiring cribbing; but on large work, sheet-steel piling should be used.

When it is necessary to operate a continuous pump, the footing concrete should be mixed and placed dry; thus it may be placed when the trench is filled with water.

Securing Dry Cellars. In localities where a porous or sandy soil exists to the depth of six or more feet, cellars are usually dry without the use of any preventive to dampness; but where compact soil exists, usually about 80 per cent of all present cellars are more or less subject to dampness, as few have been waterproofed. That concrete, like brick and stone, is a conductor of dampness is known; but that it is more readily adapted to waterproofing, only those experienced in waterproofing walls below the grade line have appreciated.

Physicians have long realized that a large amount of sickness is caused by damp cellars; but, as waterproofing does not add to the appearance while adding to the cost, it is usually omitted, though medical bills more than make up this additional cost in a few years.

In the illustration, Fig. 17, it must be remembered that the piping shown is for drainage only, and no provisions are shown for sewerage plumbing, which will require separate piping and should never be connected to the drainage sewer nearer to the building than beyond the last trap shown.

A monolithic concrete wall below grade is the cheapest and strongest; and when water-

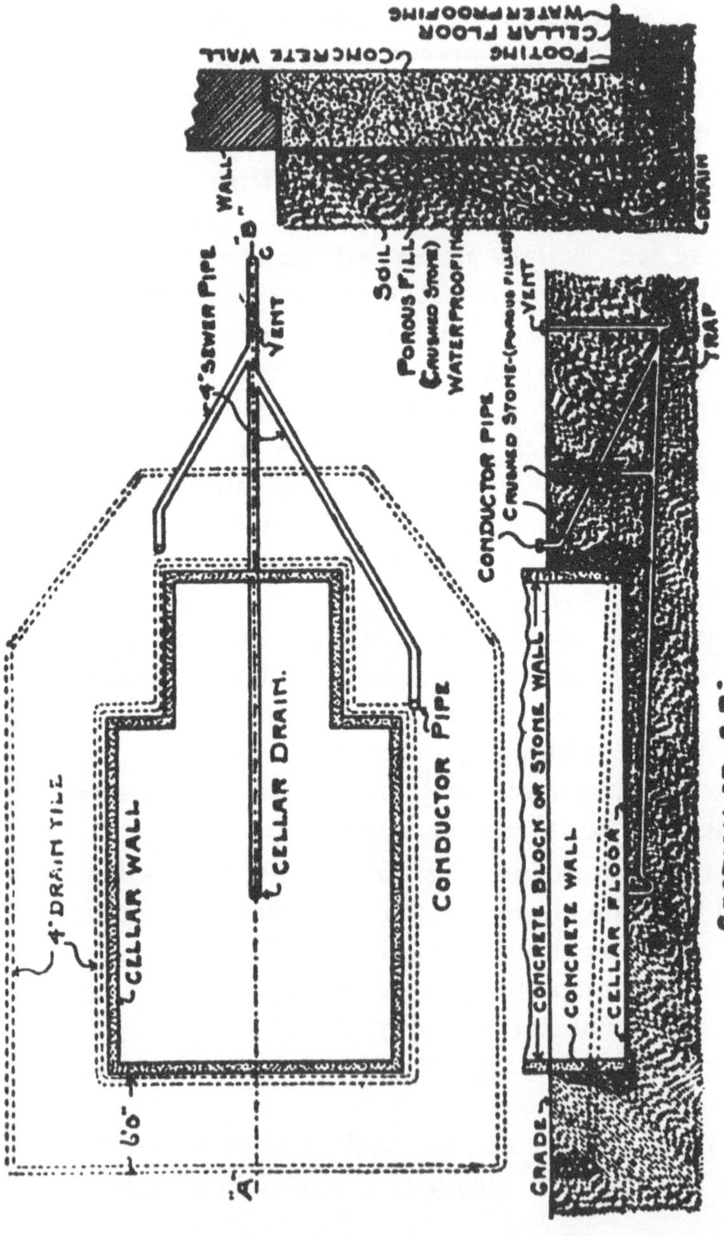

Fig. 17. Waterproofed Cellar with Drainage Piping.

proofed on the outside and on the top with the offset shown, with any positive waterproofing, it will insure dry walls. It, however, causes water to remain on the outside, which is also injurious to health; and nothing but proper drainage will overcome this evil.

Perhaps the best method of securing the necessary drainage consists in loosely placing crushed rock against the wall, with a four- or six-inch porous drain-tile, joints not cemented, placed in the bottom of the trench. The drain-tile must have no less than one foot fall or drop in twenty feet. The size of the drain-tile depends upon the length of wall, and four-inch is sufficient for buildings less than sixty feet long.

In localities where clay soil or hardpan is found, it is necessary to place another drain six feet from the building wall, which is placed in a trench of sufficient depth to be free from frost; this drain is also covered with crushed stone or brickbats, allowing space to cover with soil of sufficient depth to insure proper nourishment for the lawn. There are numerous materials that can be be used for the porous fill, crushed sandstone or brickbats being perhaps the best; but gravel or coarse cinders are acceptable.

In no instance should any part of the drain nearest the wall be above the cellar floor level, but it may be much lower, the outside or lawn drain depth being governed by frost depth.

The cellar and conductor drains should be

made of socket sewer-pipe well cemented at the joints, and have a trap at every opening on the inside of the building, and one trap after all connecting drains have been entered into the outlet; and this trap must have a vent-pipe to prevent the formation of noxious gases. Some contend that the conductor pipe having an iron pipe from the grade to the eaves of the roof, makes the best vent possible; but drains connected with street sewers often carry gases from the sewers when the traps are not water-sealed, in which event the conductor would be an outlet for such gases.

Cement Cellar Floors. No matter what kind of foundation walls are used, the floor of the up-to-date basement or cellar is of concrete. The construction is very similar to that for cement sidewalks. No sub-foundation is, however, necessary as a general thing. Level and pack the earth surface and lay down 4 inches of concrete. Float smooth, giving all sections a slight slope toward some common drain point. When the concrete has become slightly hardened, apply a half-inch top dressing of neat cement, or a rich cement mortar. This dressing should be rounded up in the corners and made continuous with the side wall finish.

To provide for cleaning water, or for any other moisture that may get in at any time, a tile drain leading outside the basement wall should be provided.

Model Specification for the Cement Work for an Average House

The following is that part of the specifications which applies to the cement and concrete work on the average residence as made use of by one of the leading Chicago architectural establishments.

FOOTINGS. All footings under all main walls, chimneys, and piers, to be of concrete, composed of 1 part Portland cement, 3 parts sand, and 4 parts crushed stone; mix well, dry; and then add enough water so that the concrete can be well tamped into place. Use best Portland cement, and leave to set in one solid mass.

CONCRETE WORK. The concrete contractor, before proceeding with the work, must see that all trenches and other excavations are of the proper depth and dimensions, as called for on the drawings, to properly receive the concrete, which in all cases must go down to a solid foundation.

The contractor must furnish and provide all labor, cement, sand, gravel, stone, water, and other necessary materials; also all tools, machines, forms, planking, and all structural iron work that may be required to carry out the work according to the plans and these specifications; and the contractor will not be allowed to make any changes without an order signed by the superintendent.

All concrete work to be composed of 1 part fresh Portland cement, delivered in labeled sacks; 3 parts clean, coarse sand; and 4 parts crushed stone or clean, coarse gravel, all thoroughly mixed and tamped in place with iron tampers.

All concrete must be used as soon as mixed; and should any concrete start to set before it is in its place, the same must not be used in any way. All concrete walls and piers must be run up true, plumb, square, and level, and must be complete and perfect in every respect. Foundation walls to be plastered with cement on inside, and waterproofed on outside.

CEMENT FLOORS. Entire basement floor surface and cellar entrance to be cemented as follows: first make pole 5 inches longer than height of basement when completed, and level off all clay, so that pole can be set straight at any point in basement under joist; and if any parts are found much lower, same must be filled in with clay, tamped solid and brought to a true level with clay surface. Over this surface apply 4 inches of grouting, composed of 1 part pure Portland cement, 2 parts clean, sharp sand, and 4 parts small, clean, broken stone, or gravel, to be mixed as follows: Place the gravel in mixing box, then cover with the sand, placing the cement on top; sprinkle the whole with hose, and mix thoroughly and as quickly as possible with shovels, taking the cement to the place where it is to be laid and spreading it in layers of uniform thickness, afterwards tamping it thoroughly with heavy tamping irons. The surface must be brought to a true level and then left to set hard. Any concrete that is broken or injured after commencing to set must be immediately removed and replaced by new, composed of new materials.

CEMENT TOPPING. Top surface to be ¼ inch thick, composed of 1 part Portland cement to 3 parts clean, sharp sand, thoroughly mixed dry, then tamped properly with water, and spread to form a true and level surface, which is to be troweled smooth and hard; before laying the top surface of concrete, the top surface of grouting must be clean swept and well wetted in order that the topping may adhere to the concrete thoroughly. Slightly pitch to floor drain.

Veranda Floors. Concrete floors for verandas, porches, and even balconies are much to be desired. While they cost a little more than wood, they are imperishable. Such floors may be built on the ground the same as walks, or on rough wooden supports. One part of Portland cement, three parts of sharp sand, and three to five parts of gravel, mixed with sufficient water to become plastic, and tamped in position to a thickness of 3 inches, is the rule. The wearing coat is a half-inch thick, and made of 1 part Portland cement and 2 parts sand. The top coat should be smoothed with a plasterer's trowel, and allowed to stand for six to eight days, being kept damp.

Steps for Residences. The growing demand for concrete steps for residence entrances has led to the construction of the type of step shown in Fig. 18. These steps or treads are made at the hollow block plant and kept in stock. The mould, a cross-section of which is shown, is inexpensive and allows the use of rich mix for facing, and wire cloth or expanded metal for reinforcing, being practically a face-down mould with a round corner on the skids for tilting up before removing the pallet.

The tread and riser are adapted to any height of riser, as a study of the section steps shows. Should a narrow tread be desired, it is easy to trim down the pointed edge projection against the riser above.

The cost of these treads is less than solid ones, there being a saving of half the concrete, a

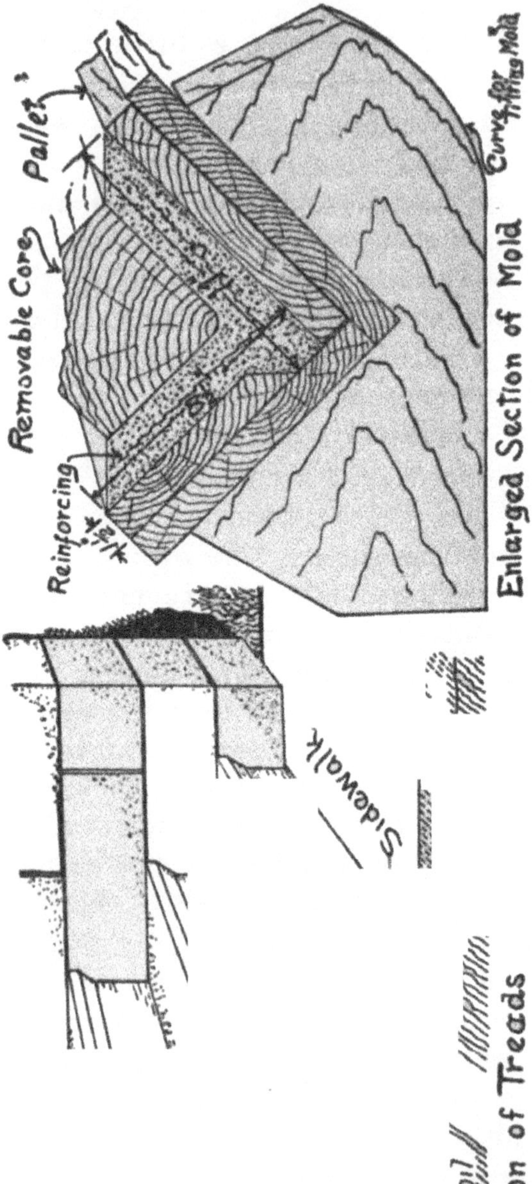

Fig. 18. Construction of Concrete Steps for Residence.

saving of labor in applying the richer facing, and only two-fifths the weight to handle.

Only one mould is necessary, but the removable core should consist of several short blocks, so that any length can be met.

The pallets should be of sufficient length for the longest steps required, and the ends of the mould sliding over the pallet permits short lengths being made on long pallets.

When return ends are desired, a special end board is necessary, which, with an extension on the end of core, admits making a full end of step, thus making the supporting wall at end of steps unnecessary.

Only two-inch surfaced lumber should be used for making the mould, and the skids should not be more than three feet apart. The removable core need not be solid as shown; but this is preferable, for, by tamping this core after the mould is filled, the concrete becomes very compact.

Round rods can be used for reinforcing, but wire cloth or expanded metal lathing is preferable, being more binding and nearly always cheaper.

The supports for these steps are the same as for stone—namely, suporting walls under each end, also a center wall under center for lengths of six feet and over.

A CONCRETE FIREPLACE

A special formula for the blocks used in the

construction of a fireplace is necessary to guard against discolorations from the heat. Blocks made in the following proportions will prove satisfactory for the fireplace and chimney construction: One part Portland cement burned at 900 degress or more, one part fine sand, three parts coarse sand. Blocks must be well tamped, and kept moist for a week. After the blocks are laid, cover all surfaces exposed to the fire with a solution of one-half pound of soap, one pound of slaked lime, applied with a flat brush. This solution will fill the pores and prevent discoloration.

Concrete for Hearths. The ordinary method of construction for work of this kind is to build a brick arch for the foundation, upon which the concrete is placed. This insures perfect fire protection. The top may be dressed with a trowel, or it may be finished with tile, as some prefer. The concrete for a hearth should consist of 1 part Portland cement, 3 parts sand, and 5 parts crushed stone or gravel, all thoroughly mixed.

CONCRETE AS A SUB-FLOOR

It is often necessary to lay a tile or mosaic floor in one or more rooms of a building where wooden joists are used, such as vestibule, hall, or bathroom. Fig. 19 shows the best method of doing this, using a waterproof sub-floor of concrete. This detail shows also the proper way of joining an ordinary wood floor to the tile floor, such as in a doorway where no threshold is desired, or around the hearth of a fireplace. If the joist under this joint is a trimmer, it should be doubled. The tops of all the joists should be

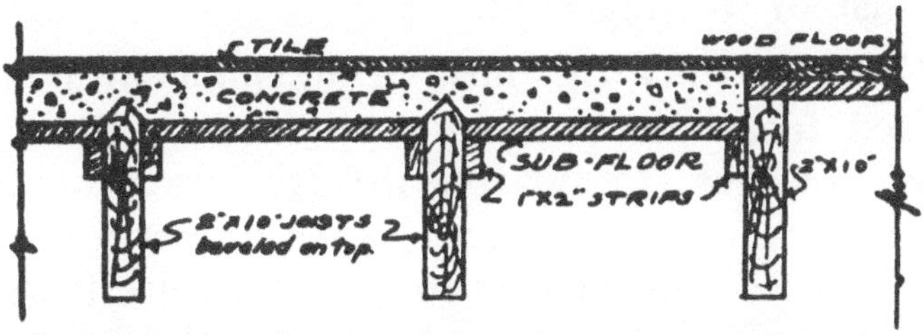

Fig. 19. Concrete Base for Tile Floor in Frame House.

beveled as shown in order to prevent cracking of the concrete above the joist. When necessary to lay this kind of floor in old buildings, the wood floor should be taken up and the floor then put in as shown in the drawing.

In factories, hotels, office buildings, etc., where finished wood floors are laid on concrete, beveled wood sleepers are used as nailing strips. These sleepers are about 2 by 3 inches in size, and are placed usually 16 inches on centers. Between the sleepers a filling of weak cinder concrete is used to hold them in place.

CEMENT PLASTER OR STUCCO WORK

One of the most popular forms of suburban house built within the past twenty years is what is called the stucco house. Three reasons may be given for its extensive use in suburban residential work: First, its cheapness; second, that it is applicable as an exterior finish to every kind of house, no matter what material is used in the construction of the walls; third, any kind of an architectural effect desired can be produced by the use of this material.

Unfortunately, this form of finish has been criticised from the standpoint of lasting quality; due, however, to failures directly traceable to poor workmanship. Stucco, when properly made and applied, will endure for an indefinite length of time. The manner of mixing, the proportion of parts, the coloring, the application and care of the walls after the plaster has been applied, make of it a problem which requires expert skill in handling. An inexperienced workman—unless he gives the matter the utmost careful study, and acquaints himself thoroughly with the methods of approved practice—will be certain to come to grief, causing regret to the owner, and creating prejudice against cement as a siding material.

The effects which may be obtained are various and interesting. Cement siding may be colored or left natural. It may be finished smooth like the ordinary sand finish plaster, or it may be stippled. Rough-cast finish is obtained by throwing pebbles mixed with thin cement upon the wall before it has had time to harden thoroughly. Cement siding may cover the house entirely; or it may be combined with wood, brick, or stone to form the wall. A very popular effect is obtained by using wood siding for the lower, and cement plaster for the upper part of the house.

Artistic effects in English half-timbered houses are due to the ease with which the spaces may be proportioned and arranged. There is an added advantage in the half-timber, in that the material in the smaller spaces is not likely to check from expansion and contraction.

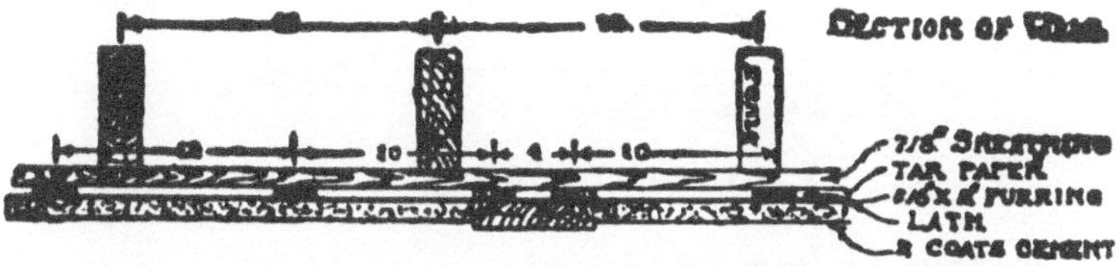

Fig. 20. Wall Section, Plastered House, Half-Timber Effect.

Framing for Cement Plaster Houses. The construction of the frame for a cement exterior differs but slightly from that for wood siding. Usually the sheathing is put on the outside of the studs, as shown in Figs. 20 and 21. Upon this is tacked tar building paper. The furring comes next. The strips are of various sizes, ⅝x1 inch, 9 inch O. C. being one of the lighter forms; they are nailed vertically, irrespective of the position of the studding. Thicker furring is used when more air space is desired than can be obtained with the thinner strips.

It is essential that the casings, cornice, base, and beltings be so made that the plaster shall be keyed to it. Strips of wood for the English half-timber effect are beveled on their edges as indicated in Figs. 20 and 21. Casings may be similarly beveled on their outer edges, except the head, which is tinned so as to turn the water. A more common method of making casings is to run a moulding entirely around the casing, al-

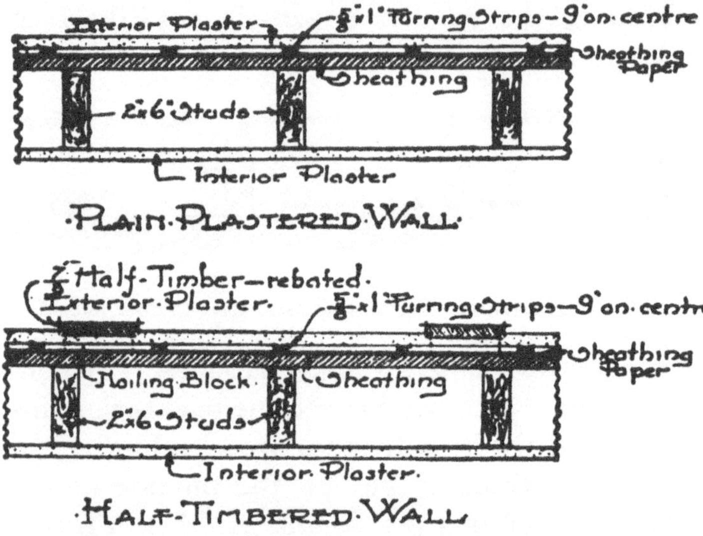

Fig. 21. Wall Sections—Cement Plastered House.

lowing it to project over the outer edge about five-eighths of an inch. Such casings have an "apron" similar to that used on the inside.

Window and Door Frames. Window framing for cement plaster houses should be set as detailed in Fig. 22, which shows the use of expanded metal lath secured to strips. This gives a better clinch for the mortar than if the lath were stapled directly to the sheathing; besides, it creates an air space and provides a wider jamb at the windows, which is essential where large plate glass is used, necessitating heavier sash than for the common double-strength glass. It is a good idea to plow or groove out the corner of the frame so that the mortar will extend under the edge of the frame. The flashing of the caps may be put on in the usual way, and plastered over. Of course it would be much easier, so far

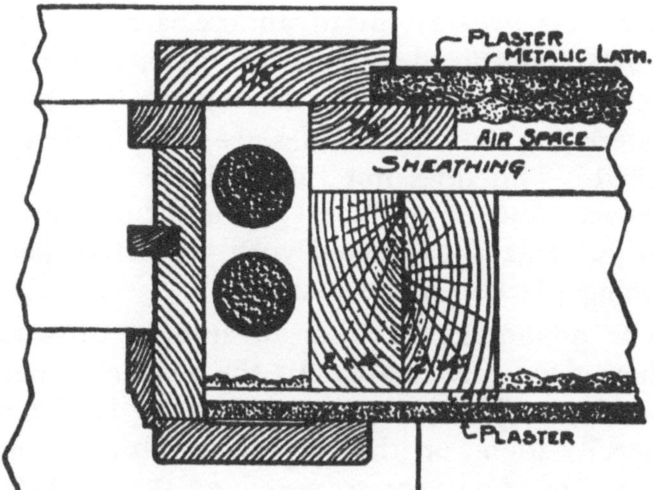

Fig. 22. Framing for Windows—Cement Plaster House.

as the plastering is concerned, to set the frames after the plastering is done; but this would not make so tight a job, especially as to the prevention of leakage at the top. The framework should be very substantial; otherwise settlement or vibration will crack the plastering.

Wood or Metal Lath for Stucco Work. The question of the relative merits of metal and wood lath is one that does not seem to be fully settled. In fact, both metal and wood have their advantages and disadvantages. Time will tell. At present both are used in about equal proportion, each having advocates with very decided opinions.

The advantage most frequently urged in behalf of metal lath is its rendering the wall fireproof. Its greatest disadvantage is its liability to rust. This disadvantage, it is claimed by manufacturers, is overcome by having the lath back-plastered so that the meshes are completely embedded. This does not fully protect the metal, however; and to overcome the difficulty, metal lath galvanized or coated with protective paint is being placed on the market.

The advantages and disadvantages of wood lath are too well known to the reader to need repeating. The poor quality of the lath now generally found on the market, which is becoming poorer from year to year, and their liability to shrink, warp, and buckle, render them far from ideal. Their cheapness has served to keep wood lath to the front in the outlying districts where fireproofing is not so much insisted upon.

Cement Plaster—How to Mix and Apply

The best results are obtained with a mixture of lime-mortar and cement. Lime-mortar alone is not durable, while cement alone is liable to show hair-line cracks, and on account of quick setting is not so readily worked; but a mixture of the two produces a permanent finish and is easily applied. The character of the work governs to some extent the proportion of cement and lime-mortar to be used. Slake good double-strength lime in plenty of water, and stir only enough to prevent the large lumps from burning. After being allowed to stand for a week or ten days, it is ready for use. The cement is then added.

The lime is usually slaked in a box raised slightly from the ground, one end being a little lower than the other, so that the slaked lime will run off. The lower end should have a sliding door, with the opening covered with a coarse wire screen which acts as a strainer, thus preventing any unslaked lumps from leaving the box. After the lime has completely slaked, it is run off into a crater of sand and the lime and sand are thoroughly mixed.

The cement is added just before the plaster is applied, and only a small quantity should be prepared at a time. The cement acts rapidly, and if this occurs before application it will lose its valne.

The cement is first thoroughly mixed with sand, and then this dry mixture of cement and sand is worked into the lime paste. The proportions by volumes should be one part lime-mortar to six parts of cement and sand mortar, the latter mixed about 1 to 1. While the first coat is wet, it is scratched deeply over the entire surface; and as soon as it has sufficiently set to support the second coat, the second coat is applied.

Finishing Coats

If a thick plaster coat is desired, a third coat is applied when the second coat is dry, but usually the finish is obtained from the second coat. The finish may be either smooth, rough-cast, pebble-dash, or slap-dash. A rough-cast finish can be obtained by using trowels covered with carpet or burlap. It is not well to trowel the surface too much, as the plaster is liable to crack and fall off if the cement is disturbed after it has started to set.

Rough-Cast. To obtain the best results, a slight excess of sand is used, and the plaster should not be very wet. The sand should be large-grained and coarse, as this adds to the rough appearance of the surface. A beautiful white finish can be secured by using white sand and white cement, or by using the white sand with ordinary Portland cement. Lime paste also whitens the plaster.

Slap-Dash. A slap-dash finish can be secured by throwing on the second coat with a wooden paddle, but it takes an expert to do good work in this manner.

Pebble-Dash. A pebble-dash surface can be secured by applying the second coat fairly wet and then throwing clean pebbles into the fresh plaster. The pebbles should be about ½ inch in diameter and should run uniformly. Before throwing them on the fresh plaster, the pebbles should be wet. The work should be started at the top and the pebbles thrown with a sweeping motion such as is used in sowing grain. The pebbles must, of course, be distributed uniformly over the surface, and must be thrown against the fresh, soft plaster with sufficient force to embed them securely.

Care must be taken not to disturb the cement after it has started to set, and in order to avoid this, the surface must be covered with the pebbles immediately after the fresh plaster is applied. A plasterer by ordinarily quick work can cover a surface 6 feet square with plaster, and then apply the pebbles.

Particular care must be taken to make the whole surface continuous; that is, one patch of plaster must not be allowed to dry before the adjoining space is covered. If this precaution is not observed, cracks are likely to occur in the finished surface.

A similar finish can be obtained by mixing the cement and lime paste to the consistency of thick cream and then adding the washed pebbles in the proportion of 5 parts plaster to 1 part pebbles by volume. This mixture is applied either with a trowel or by dashing it on with a wooden paddle about 6 inches wide. The finish is often tinted yellow, and for this purpose French yellow ochre is best. Enough coloring matter is added to the mixture to give the desired tint before applying. A yellow tint can also be secured by using brown sand or gravel.

The finished surfaces should be protected for at least two weeks with canvas curtains or bagging saturated with water.

Trouble Makers. Defects are liable to appear on cement plastered walls, (1) if too much cement is used; (2) if not applied with sufficient moisture; (3) if not troweled sufficiently; (4) if not protected from variations in temperature and draughts of air.

It may be added that improper gauging of cement and lime often causes an uneven color. Experienced plasterers overcome this easily. One who has done much of this says he thins down his lime putty so that it is so watery as to be used in mixing the cement.

Table XIV shows the area which can be cov-

TABLE XIV

Area Covered by Mortar

Mortar Produced from One Barrel of Portland Cement Mortar (3.8 cu. ft. Cement Paste.—No Lime)

Composition of Mortar	Thickness of Coat	Area Covered
1 Cement, 1 Sand	1 inch ¾ " ½ "	67 sq. ft. 90 " " 134 " "
1 Cement, 2 Sand	1 inch ¾ " ½ "	104 sq. ft. 139 " " 208 " "
1 Cement, 3 Sand	1 inch ¾ " ½ "	140 sq. ft. 187 " " 280 " "

ered by one barrel of Portland cement mortar of various mixtures, with coats of various thicknesses.

The **color effects** obtained with cement are many and are beautiful. Most of these effects are obtained, however, not as might be supposed, by mixing the dry colors in the cement, but by painting the cement after it has become dry and hard. There are two very good reasons for not mixing the colors in the cement. First, it is almost impossible to mix the mass so that it will dry with an even or uniform color. Second, most coloring matters weaken the cement. No coloring matter containing acids or anything that will act upon the alkalies in the cement, can be used; and vegetable or oil colors impair the strength of the cement.

Table XV indicates the mineral coloring materials which may be used for giving various colors and tints to cement mortar, and the proportions of coloring matter to cement:

TABLE XV

Materials Used in Coloring Mortars

Color	Mineral	Pounds Color to 100 Pounds Cement	Pounds Color to Barrel of Cement
Gray	Germantown Lamp Black	½ to ⅕	2
Black	Manganese Dioxide	12	48
Black	Excelsior Carbon Black	2	00
Blue	Ultramarine	5 to 6	30
Green	Ultramarine Green	6	24
Red	Iron Oxide	6 to 10	24
Bright Red	Pompeian or English Red	6	24
Sandstone	Red-Purple Oxide of Iron	6	24
Violet	Violet Oxide of Iron	6	24
Brown	Roasted Iron Oxide or Brown Ocher	6	24
Yellow or Buff	Yellow Ocher	6 to 10	24

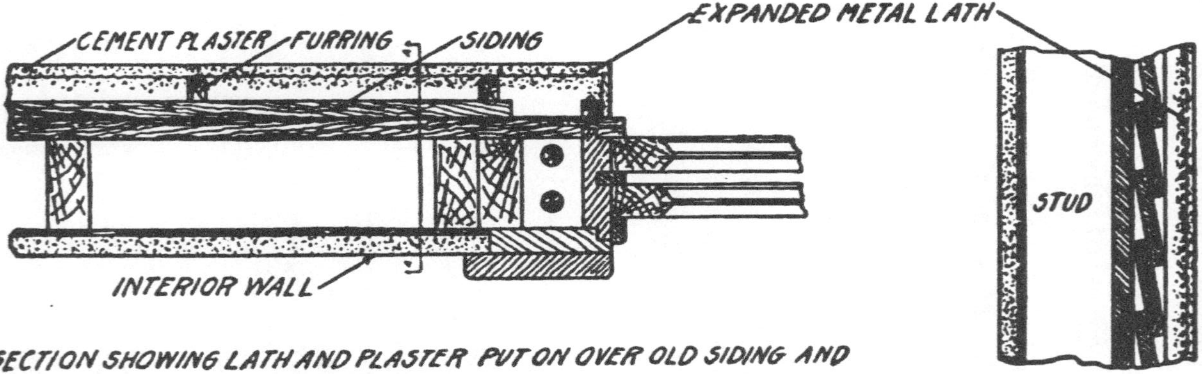

SECTION SHOWING LATH AND PLASTER PUT ON OVER OLD SIDING AND TRIM REMOVED

SECTION "AA"

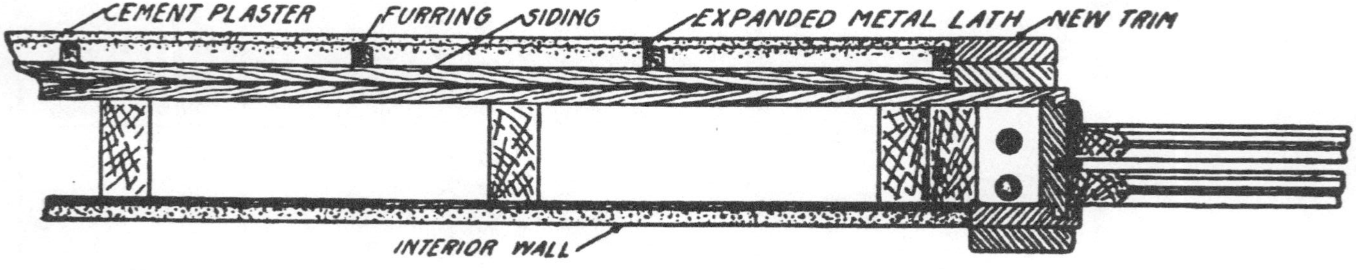

SECTION SHOWING LATH AND PLASTER PUT ON OVER SIDING AND NEW TRIM ADDED

Application of Stucco to Stone or Brick

When it is desired to cover a stone exterior with cement plaster, the surface must be thoroughly cleaned of all loose mortar and disintegrated stone; and before the plaster is applied, the surface must be thoroughly wet. The amount of wetting necessary depends upon the character of the stone of which the house is built. If it be a soft, porous stone, a great deal of water must be applied; if it be a hard, compact stone, not so much. In every case the old surface must be sufficiently saturated so that no water will be absorbed from the plaster.

Brickwork can be successfully over-plastered in the same way. The mortar joints should be raked and cleaned out to a depth of from ½ to ¾ of an inch, the brick wall thoroughly dampened, and the cement plaster then applied. The first coat should be put on with a good deal of force, driving the plaster well back into the raked joints.

Cement Plaster for "Overcoating." Cement plaster "overcoating" has been employed recently for the restoring and remodeling of many old frame dwellings which seemed to have reached the very last stage of their usefulness.

The method is simplicity itself. When the clapboarding, shingles, or other outside wood work has become too much weatherbeaten to longer keep the house trim and snug, or when for any reason it is desired to renew the exterior, metal lath is nailed right on, and the cement plaster, in two coats, is applied to it. This plaster coat is thoroughly weatherproof and waterproof; and the lime in it, acting as a preservative for wood, stops all further decay of the timbers underneath. This is all done at a cost

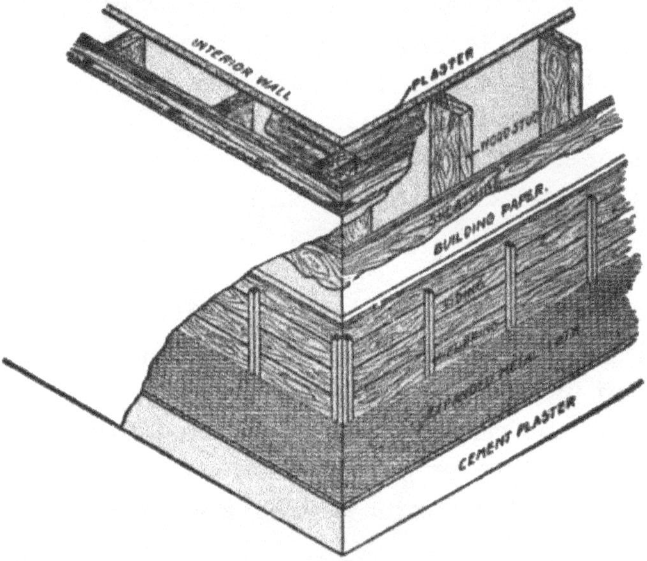

Fig. 24. Detail of Construction at Corner—Cement Plaster "Overcoating."

hardly more than double the expense of a good job of painting.

All the various finishes in use for ordinary cement surfacing are equally well adapted to this overcoating process.

If no alterations of plan or design are intended, the method of procedure may be as mentioned above, for the most inexpensive job; or it may be as indicated in the accompanying drawings, Figs. 23 and 24. In the latter case, it will be necessary only to rip off the window and door trim, and, after nailing wood or metal furring vertically to the weather boarding on 12-inch centers, attach metal lath outside of the furring to cover all sides of the house. If extra warmth is desired, it has been suggested that building paper could be first applied over the siding and under the furring strips.

The cost of such cement plaster work complete, including lath, furring strips, and sheathing paper, ranges from 8 to 12 cents per square foot, including contractor's profit. This cost applies to the straight run of wall; extras, of course, should be added for framing around doors, windows, copings, cornices, etc.

In some cases it has seemed wise to remove the shingles or siding before the overcoating is done. In one such job of this kind the total cost for labor and material was 62 cents per square yard.

In considering this expenditure in the light of an investment, it has been the experience of many that the saving in fuel and repairs otherwise necessary has been equal to from 10 to 20 per cent annually on the cost of the work.

The preparation of the cement plaster and its application on such a job is practically the same as for other stucco work. The following is a good specification for it:

Specifications for Cement Stucco

MATERIALS. The materials composing the stucco shall consist of:

(a) Portland cement which has been carefully tested and found to satisfactorily meet the requirements of the specification of the American Society for Testing Materials.

(b) Sand which is practically free from organic matter and uniformly graded in size from coarse to fine. Preference shall be given to a sand of spherical grains.

(c) Hydrated lime which has been slaked with excess of water from double strength lime, and allowed to stand at least a week before being used.

PROPORTIONS. The proportions of the above specified materials by volume shall be 5 parts cement, 12 parts sand, and 1 part lime paste.

MIXING. The cement shall be thoroughly mixed with the dampened sand, and sufficient water added to give proper working consistency. The lime paste shall then be added, and the whole composition most thoroughly worked until perfectly homogeneous. This composition shall be made up only in lots that can be immediately applied, and any material that has been mixed with water over thirty minutes before applying shall be rejected.

APPLICATION. All walls shown on elevation for stucco finish shall be two-coat work. The first coat shall be prepared as specified above, with the addition of long cow hair when applied to metal lath. The face of the first coat shall be thoroughly scratched over to form a key for the finish coat, which shall be applied to a total thickness of 1 inch, when the first coat has set sufficiently hard to safely hold it. The finish coat shall be carefully floated free from any porous imperfections.

When plastering over a masonry surface, special care must be taken to saturate thoroughly with water and to apply the plaster at once.

WATERPROOFING. The stucco shall be thoroughly waterproofed. Any one of the standard compounds, at the discretion of the architect, may be used. Follow the directions furnished by the manufacturer closely.

DRYING. Spécial care shall be taken to avoid too rapid drying. If in direct rays of the sun, it shall be protected with a damp canvas or burlap, and, when sufficiently resistive, should be frequently sprinkled with water.

No exterior plastering shall be permitted until all interior partitions are studded up and completely braced.

CEMENT BLOCK CONSTRUCTION

The Block Industry. No other department of the cement industry has so felt the need of standard specifications and uniform instructions as has the manufacture of cement blocks.

There is to-day a large and growing demand for this material, and its general and almost unlimited use is retarded only by lack of confidence on the part of architects, builders, and residence owners, who see only the wretched results that attend the efforts of the misinformed and inexperienced, and overlook the splendid possibilities of this form of construction in the hands of skilled and experienced operators.

In considering the requirements that cement blocks should meet as a structural material, we must take into account the use to which they are to be put.

We have in brick classification, the terra-cotta brick, mud brick, and dry-pressed face brick, and the hard-burned, medium, and light common brick—all of which find extensive and legitimate use, and yet vary widely in strength, fireproof qualities, and appearance.

The granites, limestones, sandstones, and marbles are generally accepted in first-class construction, and yet differ greatly in weather- and fire-resisting qualities.

Lumber, of course, is very combustible; and yet the different varieties show marked contrast in strength, durability, and fire-resisting qualities, and we have still to learn of any municipal requirements stipulating the kind of lumber for building construction.

With these facts in mind, is it not fair to ask that some latitude be granted in the manufacture and use of cement blocks?

If an owner in most localities chooses to build the outside walls of his factory or residence of light-burned common brick, showing an absorption of 30 per cent water, who is there to raise objection? In fact, the average so-called hard-burned brick will absorb 20 to 22 per cent water, and will pass muster under most municipal and architects' requirements; yet our leading municipal specifications require that cement blocks shall not exceed 15 per cent absorption, regardless of the use to which they are to be put.

Uses of Cement Blocks. Cement blocks may be properly used in substitution of other materials for:

1. Foundations.
2. Exterior and superstructure walls carrying weight.

3. Curtain walls, exterior and interior.
4. Fire walls and partitions.
5. Veneering.
6. Retaining walls.
7. Cornice, trim, and ornamental work.
8. Filler blocks for floor slabs.
9. Chimney flues, etc., etc.

In this variety of work it is at once seen that uniform quality—and the highest quality—is now required.

Experience in the use of other materials has taught us to recognize, practically without repeated or preliminary tests, the quality of most materials for which cement blocks are substituted; and this fact alone gives these older materials an advantage over the newer.

Commercial, local, and natural causes, however, are calling for the more extensive use of cement blocks. This demand will increase as our manufacturers of cement blocks gain experience, and through the observance of rational building requirements. It is of prime importance to every city and town in this country, having a building code, that it should recognize and include cement blocks as a building material.

Standard Specifications for Concrete Blocks

The writer of the specifications herewith submitted, Mr. E. S. Larned, C. E., as Chairman of the Committee on Tests of Cement and Cement Products of the National Association of Cement Users, recommended in a report, that a

Specification Committee be appointed by the Association to draw up a standard specification and uniform instructions covering the manufacture of cement blocks, with the hope that this form, when prepared, might be offered to all the cities and leading towns in the United States for adoption.

As a basis upon which to consider the matter of standard specifications and uniform instructions, his suggestions included the following in part:

Cement. Only a true high-grade Portland cement meeting the requirements and tests of the standard specifications of the American Society for Testing Materials shall be used in the manufacture of cement blocks for building construction.

Unit of Measurement. The barrel of Portland cement shall weigh 380 pounds net, either in barrels or subdivisions thereof made up of cloth or paper bags; and a cubic foot of cement packed as received from the manufacturer shall be called 100 pounds of the equivalent of 3.8 cubic feet per barrel. Cement shall be gauged or measured either in the original package as received from the manufacturer, or may be weighed and so proportioned; but under no circumstances shall it be measured loose in bulk, for the reason that when so measured it increases in volume from 20 to 33 per cent, resulting in a deficiency of cement.

Sand, or the fine aggregate, shall be suitable siliceous material passing the one-fourth-inch mesh sieve, and containing not over ten per cent of clean, unobjectionable material passing the No. 100 sieve.

Only clean, sharp, and gritty sand, graduated in size from fine to coarse and free from impurities, can be depended upon for the best results. Soil, earth, clay, and fine, "dead" sand are injurious to sand, and at times extremely dangerous, particularly in dry and semi-wet mortars; and they also materially retard the hardening of the cement. An unknown or doubtful sand should be carefully tested before use, to determine its value as a mortar ingredient. Screenings from crushed trap rock, granite, hard limestone, and gravel stones are generally better than bank sand, river sand, or beach sand in Portland cement mortars (but not so when used with natural cement, unless the very fine material be excluded).

So-called clean but very fine sand has caused much trouble in cement work, and should always be avoided, or, if impossible to obtain better, the proportion of cement should be increased. Stone screenings and sharp, coarse sand may be mixed with good results; and this mixture offers some advantages, particularly in making sand-cement blocks.

For foundations or superstructure walls exposed to weather, carrying not over eight tons per square foot, the maximum proportion shall not exceed four parts sand to one part cement. This proportion, however, requires extreme care in mixing for uniform strength and will not produce water-tight blocks. We recommend for general work not over three parts sand, if well graded, to one part cement, and the further addition of from two to four parts of clean gravel stones passing the three-fourths-inch sieve and retained on a one-fourth-inch mesh sieve, or clean screened broken stone of the same sizes. These proportions, with proper materials and due care in making and curing, will produce blocks capable of offering a resistance to crushing of from 1,500 to 2,500 pounds per square inch at twenty-eight days.

For the best fireproof qualities, limestone screenings or broken sizes should be excluded, but otherwise are all right for use.

Where greater strength is desired, particularly at short periods, from two to six weeks, we recommend the proportions of 1½ to 3 parts gravel or broken stone of sizes above given. Blocks made of cement, sand, and stone are stronger, denser, and consequently more waterproof than if made of cement and sand only, and are more economical in the quantity of cement used.

Mixing. The importance of an intimate and thorough mix cannot be overestimated. The sand and cement should first be perfectly mixed dry and the water added carefully and slowly in proper proportions, and thoroughly worked into and throughout the resultant mortar. The moistened gravel or broken stone may then be added, either by spreading same uniformly over the mortar, or by spreading the mortar uniformly over the stones; and then the whole mass shall be vigorously mixed together until the coarse aggregate is thoroughly incorporated with and distributed throughout the mortar.

We recommend mechanical mixing wherever possible, but believe in the thorough mixing of cement and sand dry, before the addition of water; this insures a better distribution of the cement throughout the sand, particularly for mortar used in machine-made blocks of a semi-wet consistency.

Curing. This is a most important step in the process of manufacture. Blocks shall be kept moist by thorough and frequent sprinkling, or other suitable methods, under cover, protected from dry heat or wind currents for at least seven days. After removal from the curing shed, they shall be handled with extreme care, and at intervals of one or two days shall be thoroughly wet by hose sprinkling or other convenient methods. We recommend curing in an atmosphere thoroughly impregnated with steam. This method serves to supply needed moisture, prevents evaporation, and in some measure accelerates the hardening of the blocks.

We view with distrust, in the present knowledge of the chemistry of cement, any artificial, patented, or mysterious methods of effecting the quick hardening of cement blocks or other cement products. If such method be proposed, it should be thoroughly investigated by competent authority before use.

Time of Curing. This is also most important. In fixing the minimum time required for curing and aging blocks before use, due regard should be given to the proportions used. It is manifestly wrong in principle to require as long a period for a 1 to 2 or a 1 to 3 block as might seem necessary for a 1 to 4 or a 1 to 5 block; and it is obviously unsafe to attempt to use a block of lean proportions in as short a time as a rich mixture would gain the necessary strength.

This might be supposed to be met by fixing the minimum resistance to crushing of blocks (of all compositions); but it must be kept in mind that a very small percentage of the blocks used are tested, by reason of the expense, inconvenience, or lack of facilities.

The required minimum resistance to crushing of first-class blocks used for exterior and bearing walls should not be imposed upon blocks for minor and less important uses.

Marking Blocks. All cement blocks should be stamped (in process of making), showing name of manufacture, date (day, month, and year) made, and composition or proportions used. The place of manufacture, methods, and materials should also be open to inspection by representatives of the Building Department, the architect, engineer, or individual buyer.

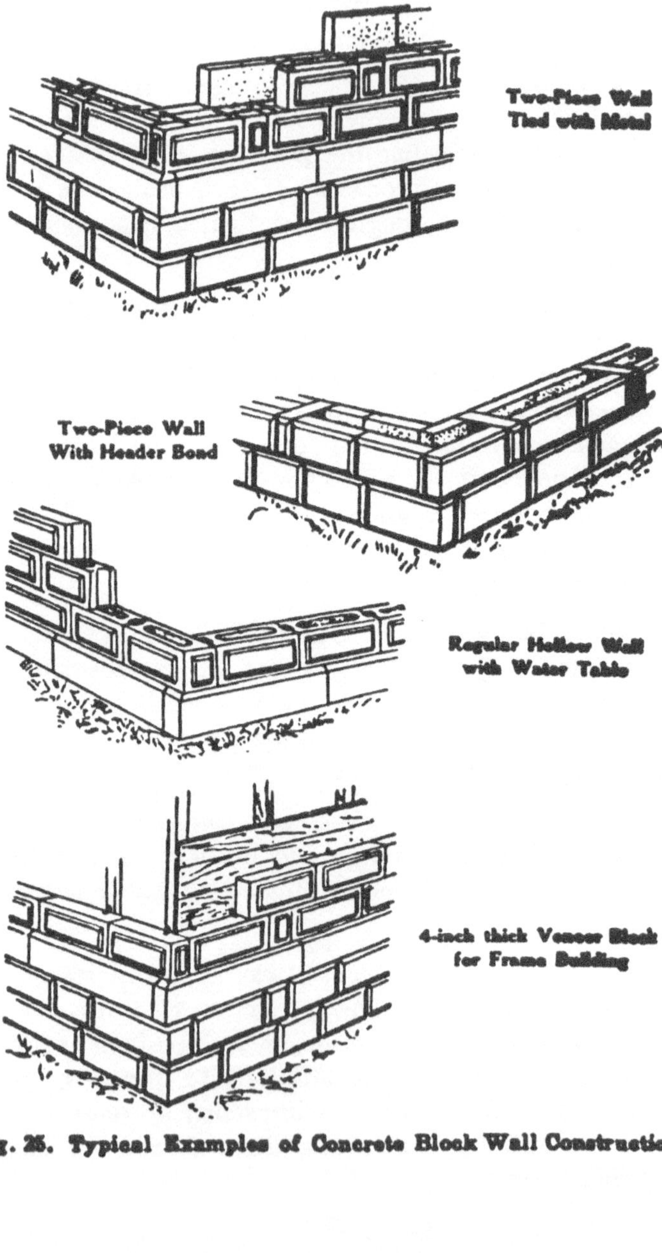

Fig. 25. Typical Examples of Concrete Block Wall Construction.

Cement Blocks in House Construction

Concrete blocks in various forms are coming to be recognized more and more as an important factor in building construction. There are comparatively few buildings erected to-day into the construction of which concrete moulded shapes, in some form or other, do not enter. Even where the building is not by any means to be classed as a block structure, we find these shapes commonly used in certain special locations. Their strength and fireproof qualities, their durability under abrasion and severe climatic conditions, their adaptability to harmonious combination with other structural materials, together with their cheapness of cost, have commended them to very common acceptance for use as **water-tables** and **belt-courses**, door and window **sills and lintels**, etc.—for which uses they have very largely supplanted the old-time cut stone.

In **ornamental shapes** for garden, lawn, and landscape decoration; as columns, capitals, pilasters, brackets, etc., for the exterior and interior embellishment even of the most elaborate structures, they are also coming into more and more extensive use.

The question of their complete function and limitations in building construction, as in the erection of houses built entirely of blocks, is one on which opinions differ considerably. The once common complaint of dampness and consequent

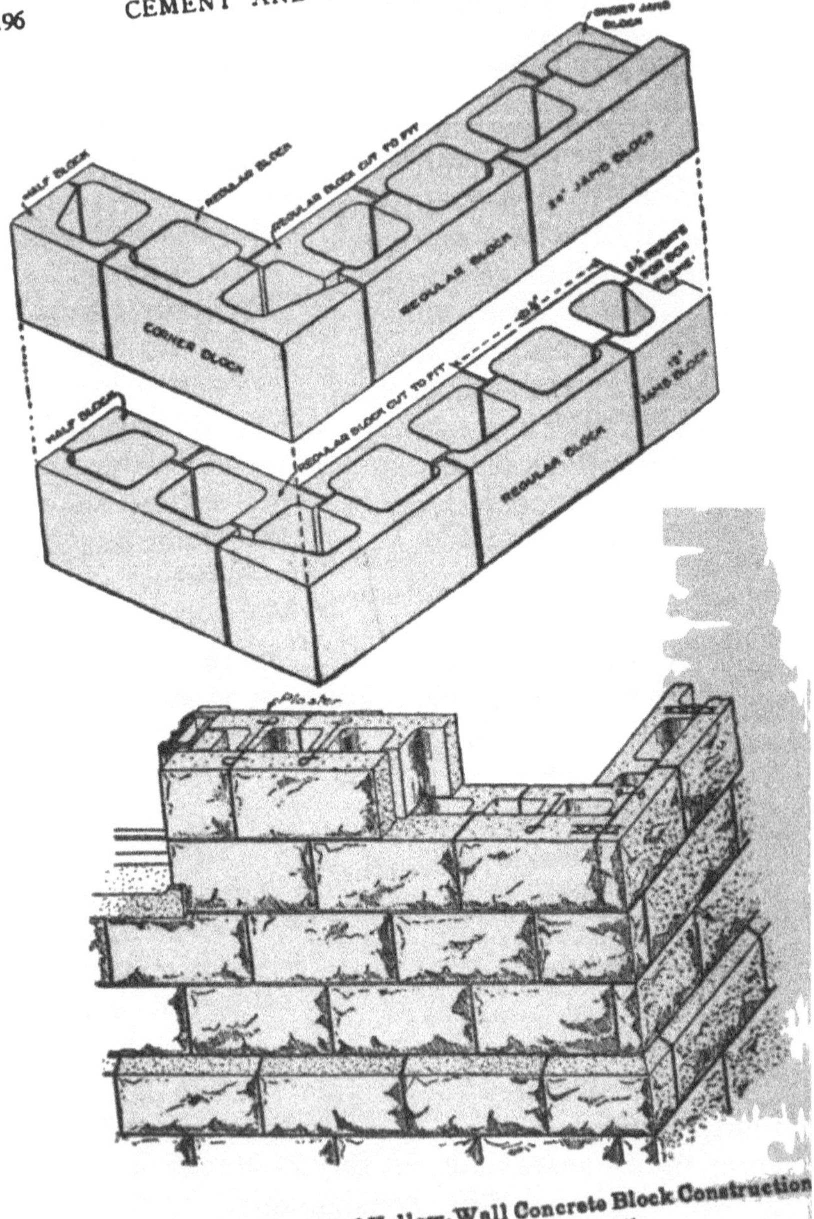

Fig. 26. Two Methods of Hollow-Wall Concrete Block Construction

Lower figure shows a veneer wall.

perfect work that marked the early, unfortunate attempts at this form of construction. But with up-to-date, improved machinery for making blocks, with our present advanced knowledge of the essential requirements in proportioning and mixing ingredients and in moulding and curing, and with modern methods available for rendering all concrete work as truly damp-proof and waterproof as work in any other material can possibly be made, there is no longer any foundation for this complaint.

There is, moreover, no doubt that with the honest avoidance of all mere attempts at imitation, with the use of appropriate faces and finishes, with due consideration of the possibilities of pleasing combination with other materials and of harmonions adaptation to surroundings in the selection of attractive outlines and color schemes, concrete blocks do lend themselves admirably to truly artistic effects in building design.

Blocks may be made solid or hollow. Some typical methods of using them in house building are illustrated in Figs. 25 and 26. Fig. 27 shows detail of window-framing for a block wall. In wall construction, solid blocks are laid in single courses for veneering, being bonded to the framework, the interior of the wall being furred, lathed, and plastered. Where the wall is built entirely of concrete blocks, these, if solid, are

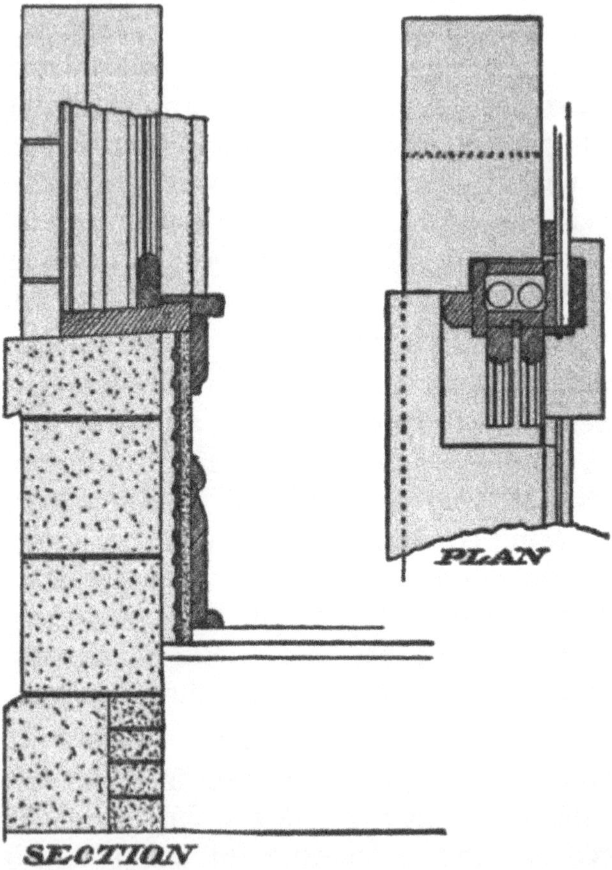

Fig. 27. Window-Framing in First-Class Block Construction.

laid in two courses, with an intervening air-space for insulation against cold and dampness, the two courses being bonded together either with metal ties or by the use of bonding headers or by means of concrete lugs on the blocks. With the hollow block, on the other hand, the insulating air-space is directly provided in the interior of the block itself; and in such cases, if proper care has been taken in the manufacture and laying of the blocks, or in the use of a reliable waterproofing process, or in both, furring and lathing may be dispensed with, and the plaster applied directly to the inner surface of the wall.

Hollow blocks are sometimes distinguished as **one-piece blocks** and **two-piece blocks.** In the "one-piece" type, the air-space represents the void that was occupied by the core around which the block was moulded. In the two-piece" type, the insulating dead air-space may be obtained by embedding pieces of metal in the block itself to tie together the outer and inner portions, or it may be formed through the shape of the pieces themselves, which, in different commercial types of blocks, are made in a variety of forms resembling more or less closely in section the steel shapes known as channels, angles, and tees, as in Fig. 26.

Number of Blocks Required. In estimating the number of blocks required, the simplest method to pursue is to figure out the superficial or face area of the building. First, multiply the length around the building (in feet) by the height of the wall (also in feet); let us call this quantity A. Find the surface of gables by multiplying width of gable by its height, and dividing by 2; call this quantity B. Find the surface of each opening by multiplying its width by its height, and add together all the areas of the openings; call this quantity C. Then add together the quantities A and B, and subtract from their sum the quantity C. The result will be the net or actual face surface to be covered with the blocks (in square feet). Divide this by 100 by simply pointing off two figures at the right, and this will give the number of squares of 100 feet each to be covered.

Then turn to Table XVI, and multiply your number of squares by the number in the right-hand column opposite the size of block to be used. The result will be the number of blocks needed for the building.

For example, suppose the superficial area to be covered is 1,525 square feet. This will be 15.25 squares of 100 feet each. If the blocks are to be 8 by 8 by 16 inches, find this size in the left-hand column of the table, and look across to the right, where you will find the number 112. Then 112×15.25=1,708, which is the number of blocks required.

Cost of Concrete Blocks. The following example of cost calculation is based on conditions that may be regarded as approximately standard:

The facing mixture is 1 part Portland cement to 2 parts coarse, sharp, clean sand; and the body of the block 1 part cement to 2 parts sand and 3 parts gravel or broken stone, the gravel or broken stone to range in size from one-fourth tu three-fourths of an inch in diameter.

One barrel cement contains 3¾ cubic feet.

One cubic yard contains 7¼ barrels.

TABLE XVI

Concrete Block Table

SIZE OF BLOCK (Height x Width x Length)	SOLID BLOCKS: Weight of Block (lbs.)	SOLID BLOCKS: No. per bbl. Cement at 1:5	SOLID BLOCKS: No. per Cu. Yd.	HOLLOW BLOCKS: Weight of Block (lbs.)	HOLLOW BLOCKS: No. per bbl. Cement at 1:5	HOLLOW BLOCKS: No. per Cu. Yd.	No. of Blocks per Square of 100 Feet
8 x 8 x 16	73	34	48	50	49	71	112
8 x 10 x 16	92	27	38	67	37	71	112
8 x 12 x 16	109	23	33	80	31	44	112
4 x 8 x 16	35	65	99	24	100	144	224
4 x 10 x 16	44	54	79	32	70	109	224
4 x 12 x 16	53	44	66	30	63	91	224
8 x 4 x 16	37	68	95	....	....	....	112
8 x 8 x 24	112	22	31	77	32	45	75
8 x 10 x 24	140	18	25	92	26	38	75
8 x 12 x 24	168	15	21	112	21	31	75
4 x 8 x 24	54	46	66	37	66	94	150
4 x 10 x 24	67	36	52	46	52	76	150
4 x 12 x 24	79	30	44	56	44	63	150
8 x 4 x 24	55	44	63	....	....	....	75

The figures given in the above table for the weight of hollow blocks and the number produced from one barrel of cement, can, of course, be taken as only approximate, since the size of the air-space in different blocks varies. It ranges from about 27 to about 32 per cent, or averages about 30 per cent, of the total space occupied by the block.

One cubic yard of sand and 3⅝ barrels of cement will give a 1:2 mixture of cement and sand.

One cubic yard of sand and gravel and 1½ barrels of cement will give a 1:5 concrete mixture.

For manufacturing 100 blocks 8 by 8 by 16 inches in size, there are needed 2.24 barrels of cement, 0.68 cubic yard of sand, and 1.06 cubic yards of gravel or broken stone. The cost of making 100 blocks will therefore be approximately as follows:

2.24 bbls. best Portland cement @ $2.00 per bbl.	= $4.48
0.68 cu. yd. sand @ $1.00 per cu. yd.	= .68
1.06 cu. yds. gravel or broken stone @ $1.50 per cu. yd.	= 1.59
Cost of labor for 100 blocks	= 1.75
Incidentals for safe margin per 100 blocks	= .50
Total cost of 100 blocks 8x8x16 in.	= $9.00

The above figures are conservative estimates of cost for materials and labor. They may vary to a greater or less degree, depending on the locality—distance from sources of supply, local labor conditions, etc.

Ordinarily blocks cost less than common brick. For comparison of cost with brick construction, the following figures showing the equivalent number of brick displaced by blocks of various sizes, will be found of help:

8x 8x16 in.	14.22
8x10x16 in.	17.77
8x12x16 in.	21.33
4x 8x16 in.	7.11
4x10x16 in.	8.88
4x12x16 in.	10.66
8x 4x16 in.	7.11
8x 8x24 in.	21.33
8x10x24 in.	26.66
8x12x24 in.	32.00
4x 8x24 in.	10.66
4x10x24 in.	13.33
4x12x24 in.	16.00
8x 4x24 in.	10.66

Cement Brick for Chimneys. Concrete blocks and cement brick make desirable chimneys; and, if care is used in the construction, entire satisfaction ought to be attained. Dry concrete being fireproof to the extent of its raw material, it has been found that sand usually endures more heat than cement. It is therefore necessary to select a cement that has been highly burned—no less than 1,200 degrees—which will make a chimney safe. But the chimney may discolor at 800 degrees without injury. For wood and soft coal fires, Portland cement is acceptable. For hard coal and coke, the cement must be selected, and gravel, limestone, and soft sandstone must be omitted. The inside of the chimney should be plastered with mortar made of one part cement to three parts of sand, mixed with strong salt water.

Cement Shingles and Roofing Tile

Roof coverings made of concrete are fast becoming popular, both on account of the artistic effects which they render possible, and because of their qualities of endurance.

Cement shingles are made in a variety of forms named in some cases from their shape, in other cases after their inventors. Some are flat, with grooved interlocking ends or sides; one has a corrugated surface, with double-grooved interlocking sides; another has tapering ribs extending to above the middle of the shingle, convex on the upper side and concave on the under side,

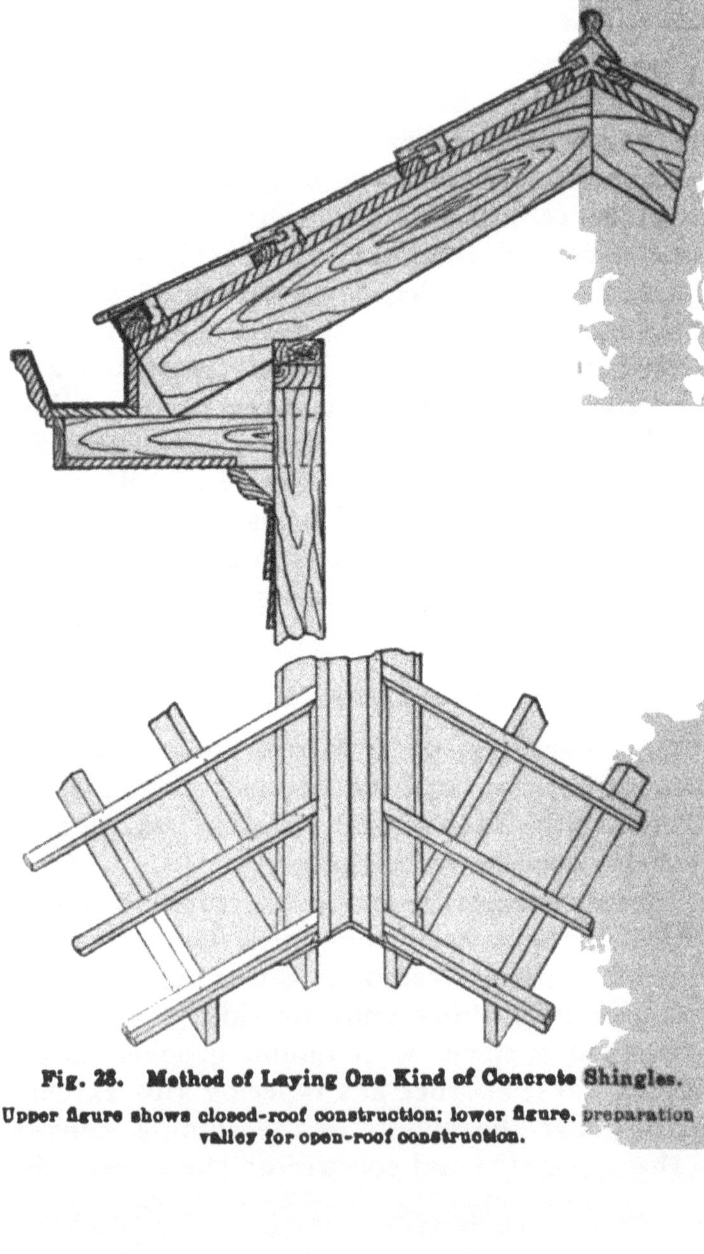

Fig. 28. Method of Laying One Kind of Concrete Shingles.

Upper figure shows closed-roof construction; lower figure, preparation valley for open-roof construction.

the shingles overlapping to the middle of the next lower course so that any water entering through the cracks between the shingles is confined to the valley between the ribs.

Different methods of laying cement shingles are adopted. In some cases they are fastened by means of nails or screws driven through perforations provided for the purpose. In other cases, small pieces of asbestos fabric are embedded in the shingles, being sufficiently flexible to allow nails to be driven through them without cracking the shingles. And in other cases each shingle is provided with a copper wire embedded near the lower end, to be securely wrapped around a nail driven in a cleat or in the roof sheathing (see Fig. 28).

SYSTEMS OF HOUSE CONSTRUCTION

Among the special systems of house construction recently evolved may be mentioned the following:

The Edison House. Thomas A. Edison, the inventor, is perfecting a system of construction which is popularly known as the **poured cement house.** By this system he intends to use cast-iron moulds which will be so arranged that the entire house, a two-story structure designed for working people, can be poured in six hours after the moulds are set up. See Plate 3.

This house is for one family, with a floor plan 25 by 30 feet. It is intended to be built on lots

40 by 60 feet, giving lawn and small garden room.

The front porch extends 8 feet, and the back porch 3 feet.

On the first floor is a large front room 14 by 23, by 9½ feet high, intended as a living room; and a kitchen in the back 14 by 20, by 9½ feet high. In the corner of the front room is a wide staircase leading to the second floor.

The second floor contains two large bedrooms, a wide hall, and a roomy bathroom (7 feet 6 inches by 7 feet 6 inches, by 8 feet 2 inches high). The third floor has two large rooms.

Each room has large windows, so that there is an abundance of light and fresh air.

The cellar, 7 feet 6 inches high, extends under the whole house, and will contain the boiler, wash-tubs, and coal bunker. The main room, as well as the outside of the house, will be richly decorated.

The decorations are cast with the house and therefore form part of the structure, instead of being stuck on, as is done at the present time.

Other Systems. Several other systems of monolithic or poured cement house construction have been developed during recent years, and have met with some success. In general the walls have been molded in slabs or in units to be afterward assembled, or the concrete has been deposited around skeleton framing systems of wood or metal.

SIDEWALK CONSTRUCTION

It may be said that cement sidewalks are now taking the place of all other kinds of walks in all cities and towns of the country. They provide an even surface for pedestrians, are permanent, and present a pleasing appearance. The cost of the concrete sidewalk makes it competitive with all other materials; but in constructing such walks the maker must have an intelligent knowledge of the process in order to assure a walk that will be satisfactory.

In all sidewalk work, Portland cement is used. In small jobs, it is only necessary to secure cement from a reputable manufacturer; but where the quantity of work will justify, it is advisable to have the cement tested.

Crushed Stone. The first requisite in the selection of the stone for sidewalk as well as other concrete work is **cleanliness.** This is absolutely essential to strength in the concrete. In selecting an aggregate, the character of the surfaces presented by the particles should always receive close attention; these must be hard and permanent. A covering of any fine material will interfere with the cement or mortar getting into contact with the surface of the aggregate, and the strength will be reduced proportionately. An excellent precaution in this respect is to avoid **the use of dirty materials.**

In order to obtain the best results, the **aggregates should be well graded**; that is, they must

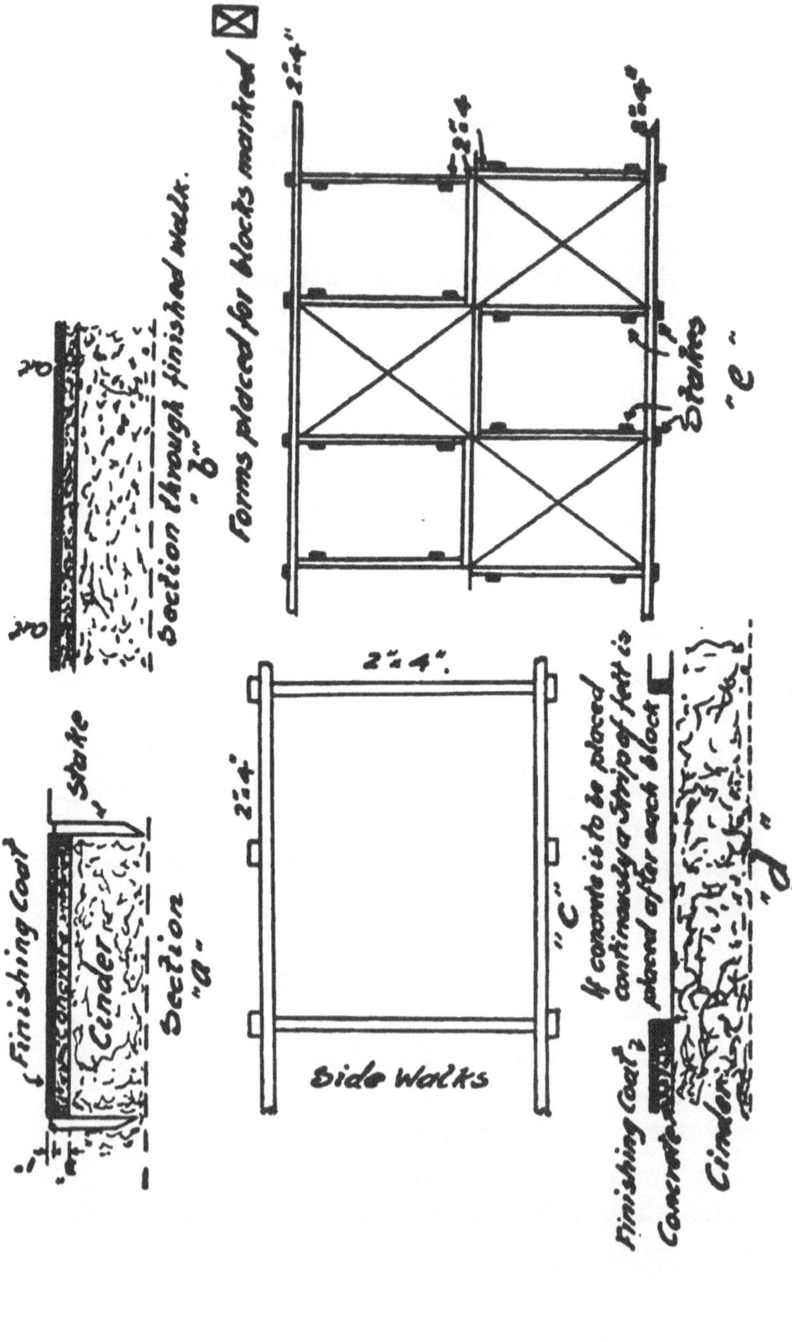

Concrete Sidewalk Construction.

not contain an excess of one-size particles, and must contain but a small percentage of fine particles. In the case of stone, the material will usually be quite satisfactory, provided the stone in itself is hard and durable and not affected by exposure to the elements, and provided it is prepared and marketed under conditions which assure its being clean and free from a covering of dust or other matter.

Sand and Gravel. In sand and gravel, one is dealing with entirely different materials, but materials probably to be preferred to stone and screenings, when selected with sufficient care. The use of sand and gravel is very popular, owing to the ease with which they are obtained in many localities. Where these materials are readily secured, they are frequently used as they come from the deposit, with little or no thought given to their fitness for the work in hand. This practice should be avoided.

The size of the sand grains and the relative proportion of grains of different size, have a very marked effect on the value of the sand. At least 75 per cent of a sand should be retained on a 40-mesh sieve, with the particles well distributed between that size and the size passing a 4-mesh sieve, with an increasing proportion on the coarser sieves. Such a sand will have much less total surface than one composed of equal proportions of particles on the several sieves. A sand made up entirely of fine particles will present a very much larger surface which must be covered with cement, than either of the sands above mentioned.

Aggregates exceeding 1¼ inches in diameter should not be used. 'Undoubtedly there are many gravels which would give good results, though containing larger sizes; but this limit is safe and the one most often applied to this class of work. The lower limit, ¼ inch, which is also the upper limit for sand and stone screenings, is almost universally accepted.

Foundation or **Sub-Base.** The foundation must provide a permanent bed for the walk, and serve as a means for disposing of water which would otherwise accumulate under the walk. In many localities, a well-constructed sub-base will offer sufficient drainage; but in some soils and under some conditions additional drainage is necessary.

Drainage. If water is allowed to accumulate in the sub-base, there is danger of the walk being heaved by frost. Therefore, in soil where the sub-base and the natural drainage cannot take care of the water, other drainage should be provided. The best means of supplying this additional drainage will depend somewhat upon the available outlets, etc. In some cases stone-filled trenches, properly placed at intervals along the walk, will provide adequate drainage, while in other cases a tile drain will be necessary.

Material. The material to be used for the foundation or sub-base of a walk must be of such a character as to withstand tamping, without crushing to the extent that it will prevent proper drainage. **Steam cinders** are commonly used for the sub-base; and if the fine material is climinated, they afford a solid foundation and provide excellent drainage.

Forms. In general, wood will be used for the forms, though thin strips of metal will be found convenient in forming curved lines. Also, the use of a metal cross-form or parting strip will be a guarantee against defects arising from imperfect joints or expansion. The cross-form should be made of ⅛-inch metal, with stiffeners of the same thickness on the ends and top. Wedges are to be driven from the outside into the ½-inch clearance space between the wooden side forms and the metal cross-form. Ready-made parting strips of special patented type are now on the market.

The wooden forms should be constructed of clean lumber free from warp, and at least 2 inches thick by about 5 inches wide. Surfaced lumber has advantages, but its use is not necessary.

In placing the side forms along the line of the walk, care should be taken to maintain a good alignment, and they should be leveled so as to conform with the finished grade.

Providing for Surface Drainage of Walk. The form nearest the street should be slightly below the inside form, thus providing a drain which will prevent water from collecting on the walk. The side forms should be securely staked, the stakes alternating on either side about every two feet. If the special metal cross-form is used, fewer stakes will answer, for when the form is keyed into position, it is rigidly fastened and holds the outside forms in their proper relative position. Wooden cross-forms need only be held in place by stakes on the opposite side from which the concrete is to be deposited. When the concrete is being placed, a shovelful or two will hold the cross-forms firmly until it is tamped into position.

When wooden cross-forms are used, the location of the joints should be definitely determined and plainly marked on the side forms before any concrete is placed. The cross-forms should be placed so that the face against which the concrete is to be packed is in line with the points indicating the position of the joints.

Providing for Expansion Joints. About every 50 feet one of the wooden cross-forms should be replaced by a metal parting strip, which should be left in the walk until it is opened to traffic, when it will be removed and the opening thus produced filled with paver's pitch or other suitable material. This forms an expansion joint, which insures the walk against cracking. This precaution is also necessary when a new walk abuts curbing or other cement or stone walk.

Surface Treatment. The surface treatment

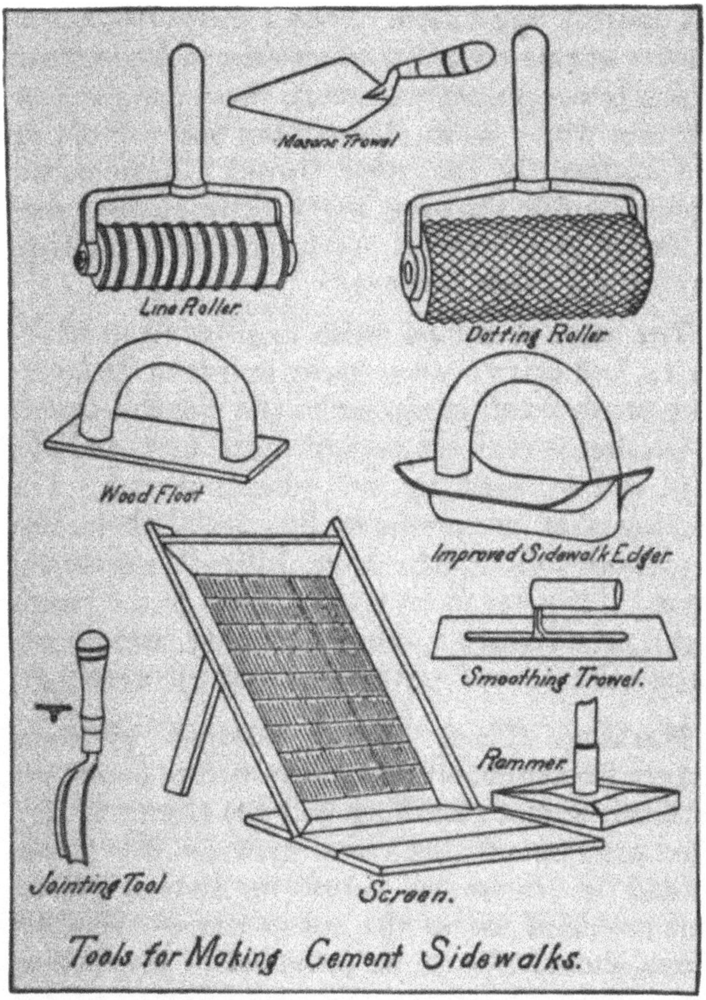

Fig. 80. Tools Especially Used in Making Concrete Sidewalks.

which a walk receives depends largely upon the practice in the community in which the work is being done. The smooth, **steel-trowel finish** is probably the most common and at the same time the poorest finish used. Such a finish frequently results in crazing or hair-checking of the surface, which is due to nothing more than a slight contraction which takes place in the film formed on the surface by the steel trowel. Besides the smooth finish showing every little blemish and variation in color, it is much more slippery than any of the other finishes.

The **wooden-trowel finish** is growing in popularity, and certainly has many points in its favor. The **brush finish** is similar to the wooden-trowel finish, but it requires an additional tool, and one that can be used for no other purpose. The finishes that are produced by special tools, like the tooth-roller, etc., have little to commend them. They are in no way superior to the rough finish produced in a simpler manner, and do not harmonize so well with the usual surroundings.

Marking. There might possibly be some chance for argument regarding surface finish, but certainly surface marking will not permit of any. The position of the joints between the blocks should be determined before the base is placed, and provided for in the construction. **Positive joints should always be provided in the base of the walk.** These are the real joints, and the markings in the top should always occur over them. It is not sufficient to make a surface marking, together with a feeble effort toward cutting through the base with a small trowel or similar instrument. **More walks are disfigured**

by failure on the part of the builder to provide proper joints than by any other cause.

Size of Block. The size and shape of the blocks into which a walk is divided are governed very largely by the width of the walk, the local practice, and personal tastes. Other points, however, should be considered; in fact, local practice and personal tastes should be eliminated entirely when walks on business streets are being constructed. Where the whole space between the building line and the curb is to be covered, many angles and irregular lines are introduced, owing to openings, steps, etc. Steps should never be constructed over a joint; nor should a joint ever be permitted to intersect a step (excepting at a joint), unless the walk and step are constructed entirely independent of each other. Joints between the blocks should be placed so as to avoid small corners and unnecessary angles; in fact, so far as possible, all blocks should be rectangular. Also the joints in new work, abutting old, should always be projected from the joints in the original work, unless a distinct opening joint is provided between the new and the old.

CURBS AND GUTTERS

Concrete curbing should be built in advance of the walk, in sectional pieces 6 feet to 8 feet long, and separated from each other and from the walk by tar paper or a cut joint, in the same manner as the walk is divided into blocks.

Curbs should be 4 inches to 7 inches wide at the top, and 5 inches to 8 inches at the bottom, with a face 6 inches to 7 inches above the gutter. The curb should stand on a concrete base 5 inches to 8 inches thick, which in turn should have a sub-base of porous material at least 12 inches thick. The gutter should be 16 inches to 20 inches broad, and 6 inches to 9 inches thick, and should also have a porous foundation at least 12 inches thick.

Keeping the above dimensions in mind, excavate a trench the combined width of the gutter and curb, and put in the sub-base of porous material. On top of this, place forms, and fill with a layer of concrete, one part Portland cement, three parts clean, coarse sand, and six parts broken stone, thick enough to fill the forms to about 3 inches below the street level. As soon as the concrete is sufficiently set to withstand pressure, place forms for the curb; and, after carefully cleaning the concrete between the forms and thoroughly wetting, fill with concrete, one part Portland cement, two and one-half parts clean, coarse sand, and five parts broken stone. When the curb has sufficiently set to withstand its own weight without bulging, remove the ¾-inch board between the face of the curb and the form (shown in Fig. 31), and with the aid of a trowel fill in the space between the concrete and the form with cement mortar, one part Portland cement and one part clean, coarse sand. The finishing coat at the top of the curb should be put on at the same time. Trowel thoroughly and smooth with a wooden float, removing face form the following day. Sprinkle often and protect from sun.

In making curbs alone, excavate 32 inches below the level of the curb, and fill with cinders, broken stone, gravel, or broken brick, to a depth

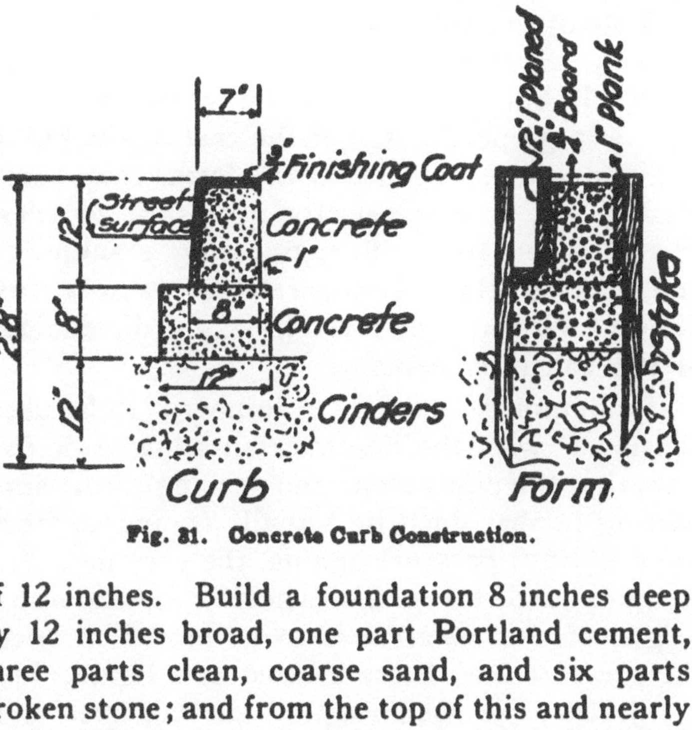

Fig. 31. Concrete Curb Construction.

of 12 inches. Build a foundation 8 inches deep by 12 inches broad, one part Portland cement, three parts clean, coarse sand, and six parts broken stone; and from the top of this and nearly flush with the rear, build a concrete wall 11¼ inches high, 7¼ inches broad at the base, and 6¼ inches at the top, the 1-inch slope to be on the face.

Remove the forms as soon as the concrete will

withstand its own weight without bulging, and put on the finishing coat in the manner as indicated above. Keep moist for several days, and protect from the sun. Measurements may be varied to suit local conditions.

DESIGN OF FORMS FOR CONCRETE WORK

"Rule-of-thumb" layout of forms in the field is being superseded by design in the drafting-room. In building construction where the forms form a large percentage of the cost of the building, and where a failure in the forms may cause loss of life, it is especially necessary to treat this question from an engineering standpoint, and many of the best concrete contractors now design their forms as carefully as the dimensions of the concrete members.

If a minimum quantity of lumber is to be used consistent with the deformation allowed, it follows that the dimensions and spacing of the supporting lumber must be actually computed from the weight or pressure against the sheeting. For columns and for walls where a considerable height of wet concrete is to be placed at once, the pressure may be calculated as a liquid. Mr. J. Douglas assumes that the concrete is a liquid of half its own weight, or 75 pounds per cubic foot.

In ordinary walls, where the concrete is placed in layers, computation is not usually necessary, since general experience has shown that maximum spacing for 1-inch boards is 2 feet, for 1½-inch plank is 4 feet, and for 2-inch plank is 5 feet. Studding generally varies from 3 by 4 inch to 4 by 6 inch, according to the character of the work and the distance between the horizontal braces or waling, 4 by 4 inch being the most useful size.

Floor forms are better based upon an allowable deflection than upon strength, in order to give sufficient stiffness to prevent partial rupture of the concrete or sagging beams.

In calculating we must add to the weight of the concrete itself—that is, to the dead load—a construction live load which may be assumed as liable to come upon the concrete while setting. Definite units of stress must also be assumed in the lumber.

We would suggest the following basis for computation, these being values which have been adopted for use:

(1) Weight of concrete, including reinforcement, 154 lbs. per cu. ft.

(2) Live load, 75 lbs. per sq. ft. upon slab, or 50 lbs. per sq. ft. in figuring beam and girder forms.

(3) For allowable compression in struts, use 600 to 1,200 lbs. per sq. in., varying with the ratio of the size of the strut to its length. (See table below). If timber beams are calculated for strength, use 750 lbs. per sq. in. extreme transverse fibre stress.

(4) Compute plank joists and timber beams by the following formula, allowing a maximum deflection of ⅛ inch:

$$d = \frac{3Wl^3}{384EI} \quad \ldots\ldots\ldots\ldots (1)$$

and,

$$I = \frac{bh^3}{12} \quad \ldots\ldots\ldots\ldots\ldots (2)$$

in which,

d=Greatest deflection in inches;
W=Total load on plank or timber;
l=Distance between supports in inches;
E=Modulus of elasticity of lumber used;
I=Moment of inertia of cross-section of plank or joist;
b=Breadth of lumber;
h=Depth of lumber.

The formula is the ordinary formula for calculating deflection, except that the coefficient is taken as an approximate mean between $\frac{1}{384}$ for a beam with fixed ends, and $\frac{5}{384}$ for a beam with ends simply supported.

For spruce lumber and other woods commonly used in form construction, E may be assumed as 1,300,000 lbs. per sq. in.

Formula (1) may be solved for I, from which the size of joist required may be readily estimated.

The weight of concrete per cubic foot is somewhat higher than is frequently used, but is none too much where a dense mixture and an ordinary percentage of steel is used. For very rough calculation, however, it is frequently convenient to remember that 144 lbs. per cubic foot is equivalent to the product of the dimensions of the beam in inches times a length of one foot.

The suggested live load is assumed to include the weight of men and barrows filled with concrete, and of structural material which may be piled upon the floor, not including, however, the weight of piles of cement or sand or stone, which should never be allowed upon a floor unless it is supported by concrete sufficiently strong to bear the weight, or by struts under all the floors below.

The units for stress in struts are somewhat higher than in timber construction, because the load is a temporary one. The extreme variation given is due to the fact that when a column or strut is longer than about sixteen times its smallest width there is a tendency to bend, which must be prevented either by bracing it both ways or by allowing a smaller load per square inch. For struts ordinarily used, the stresses given in Table XVII may be assumed for different heights.

Bracing both ways will, of course, reduce the length of a long strut.

TABLE XVII

Safe Strength of Wood Struts in Forms for Floor Construction

Pounds Per Sq. In. of Cross-Section.

Length of strut.	3" x 4"	4" x 4"	6" x 6"	8" x 8"
14 ft.		700	900	1,100
12 ft.	600	800	1,000	1,200
10 ft.	700	900	1,100	1,200
8 ft.	850	1,050	1,200	1,200
6 ft.	1,000	1,200	1,200	1,200

If the concrete floor is comparatively green, the load must be distributed by blocking, preferably of hardwood. At the top of the strut provision must be made against crushing of the wood of the plank or cross-piece. Ordinary soft wood will stand without crushing only about 700 lbs. per sq. in. across the grain; so, if the compression approaches this figure, brackets must be inserted or hardwood cleats used.

Kinds of Lumber to Use. The selection of the lumber must be governed by the character of the work and the local market. White pine is best for fine facework, and quite essential for ornamental construction cast in wooden forms. For ordinary work, however, even for the panels, white pine is apt to be too expensive; and spruce, fir, Norway pine, or the softer qualities of Southern pine must be substituted for it. Some of these woods are more liable to warp than white pine, but they are generally stiffer, and thus better adapted for struts and braces.

Kiln-dried lumber is not suitable for form construction, because of its tendency to swell when the wet concrete touches it. Very green lumber, on the other hand—especially Southern pine, which does not close up quickly when wet—may give trouble by joints opening. Therefore the middle ground—or, in other words, partially dry stuff—is usually best.

Circular Forms. In a circular form there are two sides—the inner and the outer. These may be used together, as in building a silo; or, as in a cistern, using the inner form alone; or for a column, using only the outer form. Both sides of the form are made in the same way; but the inner and outer sides cannot be made to the same pattern, as the thickness of the walls comes between the parts, making the radius of each side different.

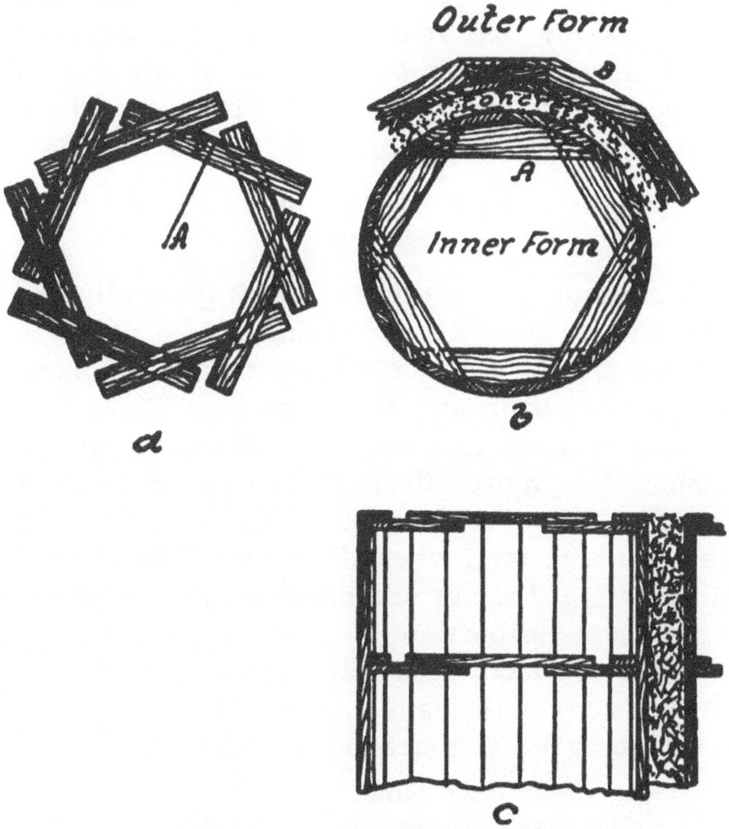

Fig. 32. Method of Constructing Circular Forms.

The simplest way to make a circular form is to draw a circle of the size of the form desired, and lay boards around the circumference of the circle as shown in Fig. 32 at a. These boards should be lightly tacked together in place, and, using the same measure, mark a circle upon them. They should then be knocked apart, and sawed out along the lines marked, the pieces being fastened securely together, as shown in diagram. After making two or more forms, place them at equal distances apart, and put on the side boards in the manner shown in the illustration. These boards are called lagging.

SURFACE FINISHES ON CONCRETE WORK

A pleasing and consistent finish generally has but little to do with the strength of a concrete structure, but it is not inconsistent with maximum strength in any structure.

Next to form or design, the character of the surface has most effect on the appearance of concrete, whether in a building, arch, wall, or abutment; in fact, when the view is had at very close range, the surface finish may take precedence over proportion.

The imperfections in the exposed surfaces of concrete are due mainly to well-known causes which may be summed up as follows:

1. Imperfectly made forms.
2. Badly mixed concrete.
3. Carelessly laid concrete.
4. Efflorescence and discoloration of the surface after the forms are removed.

Cement Plaster Finish. Where a cement finish is desired on concrete walls, the finish should be placed while the wall is being built. The rough concrete is spaded back from the forms, and the rich mortar placed in front of it. A cement finish plastered on concrete after the wall is built will usually crack and not give the best results. After the forms are removed, the concrete should be rubbed smooth and given a coat of cement wash mixed and applied as a paint.

In cases where it is necessary to plaster concrete walls, precaution should be taken to make the plaster stick. In many cases it is well to wash or scrub the surface, or to pick it to make it rough, before applying the plaster. A rich mortar of 1 part cement to from 1 to 2 parts sand should be used for such work. Lime paste added to the mortar is advisable in some cases. This increases the adhesion and lessens the liability of cracking. If a hard surface is desired, only a small amount of the paste should be used. In plaster work of this kind it is customary to brush over the surface after troweling, to remove the tension in the cement.

An inexpensive manner of plastering is what is called a **splatter-dash** coat. The mortar is thrown against the surface with a stick or paddle, making a very effective rough surface. A rough surface is generally better in appearance and less liable to crack than a smooth surface.

A method recommended as providing a good bond is to first wash the surface thoroughly with water, and then brush on a coat of neat cement. While this is wet, a coat of plaster about ¾ inch thick is put on. This is followed by succeeding coats applied about an hour apart until the desired thickness is reached. If desired, the last coat may be thrown on in order to produce a rough surface.

Plastering should as a general rule be resorted to only to fill holes and to smooth over rough places. Godfrey states that a plaster coat should either be very thin—that is, just enough to fill irregularities—or it should be 1 to 3 inches thick, so that it will have some strength in itself.

Strips of wood are often nailed into the forms to give the effect of cut stone to the surface when the forms are removed. **Dry surface finish** is produced by using a fine stone in the aggregate, mixing fairly dry, and not spading the concrete next to the forms. **Facings of brick, tile, stone,** and even **cast slabs** of concrete, are used in walls for a finish, backed up by the rougher poured work.

Brushed or Etched Finish. By varying the kind, size, and proportions of the aggregates, surface finishes of practically any desired color and texture can be obtained by brushing and washing, the possibilities being limited only by the number of different aggregates available and the combinations of same. A great variety of finishes may be produced by using red and black granite and limestone screenings, black and white marble chips and different-colored pebbles and sands, etc.

This process consists simply in scrubbing the

surface with a stiff brush, permitting it to harden for a few days, and then treating it with a dilute solution of hydrochloric acid.

Having decided upon the general color scheme and texture of the desired surface, the first step is the making and treating of small sample surfaces. The color and texture of the finished surface depend upon the color, size, and proportions of the aggregates used in the outside layer, irrespective of the composition of the wall proper, and the successful reproduction of the desired surface is dependent upon the proper selecting, grading, proportioning, and mixing of the materials, and the proper placing and finishing of the surface.

For brushed surfaces, all that is required of the forms is that the face lagging be kept true to surface and that the joints be tight. For surfaces too large to concrete in one day, the forms should be so constructed as to permit of the removal of sections of the face form.

The facing material should be from 1 to 1½ inches thick, the remaining thickness of the work being composed of ordinary concrete, but the facing and backing must be deposited at the same time so as to make one solid mass, thereby insuring perfect bond. The facing material may be applied to the forms just ahead of the backing, which is placed against and rammed into it, or the backing may be placed first and then brushed back from the form with a spade and the facing material deposited between the backing and the form. Both these methods have been successfully used.

Use of Granolithic Plates. A third and possibly the best method of placing the facing material consists of the use of what might be called a metal facing form or mould, constructed and used as follows:

To short lengths of 3/16-inch iron plates 8 or 10 inches

wide and 6 feet long, three 1 or 1½-inch angles are riveted, placing an angle at the center of the plate and one about six inches from each end. One edge of the plate should be slightly flared to assist in depositing the material, and this edge provided with handles. The metal facing plate is placed against the wall form with the handles up and the angles tight against the form. The space between it and the back of the wall filled with the concrete backing and the 1 or 1½-inch space between the metal form and the face form filled with the facing material. The metal form is drawn almost out, and after thoroughly tamping the backing against the facing the process is repeated.

Brushing and Etching. For brushed surfaces, the forms must be removed from the work as soon as possible, and the concrete surface brushed while still green. It is not possible to state how old the work should be before removing the forms and brushing the surface. This will depend upon a number of conditions—the character of the work, the cement and aggregate used, the consistency of the mixture, and very much upon the weather conditions. As a rule, in hot weather the forms can be removed the next day and the surface brushed; but in cold weather the facing form cannot be removed so soon, several days or perhaps a week being required for the concrete to attain the necessary hardness and strength. Care must be taken that the brushing is not done too soon, as little particles of aggregate will be removed, resulting in a pitted, unsightly surface. On the other hand, the longer the surface stands before being brushed the more brushing it will require to remove the film of material that has flushed to the surface. Brushing should be done just as soon as it can be without demoving particles of aggregate. When this can be done can be determined only by experimenting with the particular surface. An ordinary scrubbing brush with stiff palmetto fibres, or a metal wire brush, will answer for the work.

Two or three days after the brushing, the surface should be washed down with a dilute solution of commercial hydrochloric acid, one part acid to two parts water. The acid should be applied with an ordinary calcimining brush, and the walls thoroughly rubbed, while wet with the acid, with a stiff vegetable fibre brush. The acid should not be allowed to remain on the surface for any length of time—not over half an hour—and should be washed off with a hose and clean water. It is important that the surface should be thoroughly washed after the acid treatment, for if it is not it will have a mottled, streaky appearance.

This method of treatment removes the film of mortar that has flushed to the surface, exposes the aggregate, erases all traces of form markings, and produces a rougher, more artistic surface. The roughness of the surface breaks up the light, the color of the aggregate adds variety and life, and we have a pleasing, artistic, true concrete surface.

Bush Hammering. For wall work the surface is sometimes picked with a pointed or toothed tool. This chips off the mortar which may have flushed to the surface, and cuts away little particles of the mortar from the aggregate below. The roughening of the surface breaks up the light, gives a lighter, snappier color to the mortar itself, and, besides this, exposes the color of the aggregate below. Oftentimes where gravel is used, the stones show rusting and have various shades of browns and reds. This additional color on the concrete adds a great deal to its appearance; and when the dressing is carefully done, it gives as pleasing a surface as can be obtained economically. The dressing removes most of the traces of the form and does away with inequalities which may occur in the work.

The objection to this kind of dressing comes in the removal of the surface mortar, which is the most waterproof part of the concrete. If there is any tendency toward porosity in the mass of the concrete, it will absorb more moisture after the dressing than before it, and will accentuate the injury from frost. On well handled and properly proportioned concrete there is, however, very little danger from this, as the material is of itself very dense and waterproof.

Painting the Surface. In some localities a popular method of finish, as well as the cheapest, is to paint the surface with cement mortar. There is danger, however, in this method of marring the effect by drippings if the work is done carelessly.

There are also on the market, sold under various proprietary trade names, a multitude of special paints and finishes for concrete surfaces, giving a wide range of color effects and also serving more or less efficiently the purpose of weatherproofing and damp-proofing.

The Quimby process of wall finish is quite extensively used in Philadelphia. This method consists in using boards of such a width for the forms that each board will extend upwards through a space covered by one course of the concrete. These boards have small triangular strips of wood nailed along each edge, so that when the forms are in place two strips will come together and form a seam in the surface. As the forms are removed from work which is already in place, they are carried up, and make a new course above. As soon as the lower forms can be removed, the concrete surface is at once scrubbed with an ordinary scrubbing brush and water, and then washed off with a hose.

economy are over. These conditions led to flimsy construction; but the time has come now when the farmer can use better materials in his constructive work. All that is needed is that he shall understand the value of concrete for this purpose, its adaptability, endurance, and permanency.

The many uses to which concrete may be put by the farmer to insure to him durability and that imperishable quality so much to be desired, are being recognized more and more. Any material that gives the agriculturist the certainty that what he builds will last, not only during his lifetime, but during that of his children and his children's children, is to his advantage. And when it is considered that he can employ cement in making his improvements at a cost not exceeding that which he would have to pay if perishable lumber or timber were used, it is evident that concrete must be adopted generally by the wise home-maker and the agriculturist.

Windmill Foundation. Concrete construction will give to a windmill a solid foundation that will insure it against rotting and against being blown down.

Excavate four holes at the desired distance

apart, $2\frac{1}{2}$ feet square and 5 feet deep. Build forms for the sides, and grease them properly. Fill the forms 2 feet deep with concrete, of a jelly-like consistency, one part cement, three parts sand, and six parts gravel or broken stone, tamping well every four inches.

The holding-down bolts can be suspended from a frame over the top, care being taken to place them so that they will be in true position when the concrete is placed around them. They should be two feet long, with plates to resist the pulling strain. Fill the form with concrete flush with the top, and allow it to remain for several days before using. This will make a substantial anchorage for a steel tower. If a wooden tower is to be used, run projecting bolts up through the timber sills, and use large cast-iron washers under the bolts. The anchorage in this case should project at least six inches above the ground.

Concrete Sinks. Sinks made of concrete are considered as durable as iron. They may be made in any desired size, and reinforced with wire netting. They are cast in wood or plaster moulds. The mixture should be one part Portland cement and two parts fine crushed granite. A rebated hole should be provided for a trap. The thickness should be two to three inches.

CONCRETE TANKS AND CISTERNS

Concrete Water-Tanks. It is unquestioned that a tank built of concrete is not only more durable than one of wood, but is more sanitary. Many farmers have them installed in their barns, and tanks of this character answer the requirements in numerous manufacturing plants.

Two forms are required in the construction of a tank. One is needed for the moulding of the exterior, and a smaller one for the formation of the inner surface of the walls. Each form should be made of dressed boards without knots, which would disfigure the surface of the concrete. For a tank of medium size, say 2 feet wide, 6 feet long, and 2 feet deep, the walls should be three inches thick. Allowance must be made therefore in the making of the two forms for a space of this width.

A flat surface perfectly smooth should be provided on which to place the forms. The smaller one is put in position first, as it is the one that will form the interior walls. It would be well, before placing the larger form, to measure carefully for the placing of it, by laying it on the platform over the smaller one and marking the corners, fastening the form so that it will not move when the concrete is placed. The inner form should be fastened also for the same reason. Both forms should be greased.

The concrete mixture should consist of 1 part Portland cement, 2 parts sand, and 3 of gravel or crushed stone that will pass a quarter-inch sieve. Be careful not to make the mixture too wet. Tamping in layers of three inches is recommended for this work, each layer being followed up quickly with the succeeding one so that the fixing may be uniform. The top is last leveled off and finished. Do not disturb the forms for a week or ten days. Then they may be removed, and the drying continued.

Waterproof Concrete Tank. The suggestion is made that in all cases the water tank should if possible be slightly flaring, so that, if the water in it freezes, the pressure of the expanding ice will be less than if the sides or ends were vertical. In the case of water troughs, this shape also affords easier access.

The best mixture for a tank is 2 parts Portland cement, 3 parts sharp sand, and 5 parts gravel mixed with sufficient water to make the cement plastic or "sticky," but not thin enough to pour.

First dig the foundations, and place the footings for the wall and floor. Then mix the concrete as follows: For the foundation up to within four inches of the floor line, use 1 part Portland cement, 3 parts sharp sand, and 2 parts gravel that will pass through a one-inch sieve, and 4 parts gravel that will pass a two-inch sieve. Mix thoroughly, and ram into position. If the tank is to be as large as 30 by 40 feet and 6 feet high, this course should be covered with a 3½-inch course of 1 part Portland cement, 3 parts sand, and 3 parts gravel that will pass through a sieve of ¾-inch mesh. After this has hardened sufficiently, build the falsework with a plank on both sides, and put in all pipe connections before beginning to place the concrete for the walls. Leave all the falsework in position for at least three days, to allow the concrete to harden. After removing the plank, plaster the exterior surface with a mixture of 1 part Portland cement and 2 parts sharp sand that has

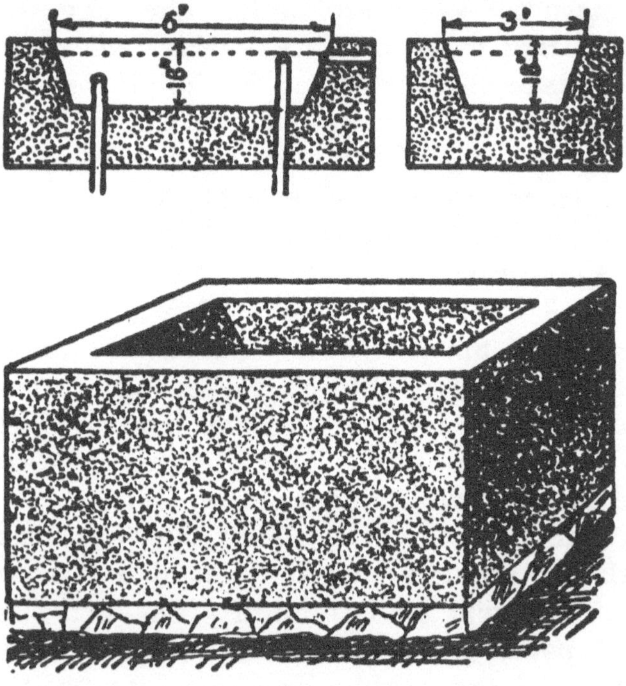

Fig. 33. A Small Concrete Tank.

passed through a screen of one-fourth-inch mesh. This will give a finished effect, and will hold if the surface is wetted before application. A waterproofing compound should be used in all the mixtures.

A trough with a solid concrete base should

be made in the same general way, the forms carried up to the desired height of trough, and the reinforcing embedded in the concrete a few inches from the inside. Troughs should be protected from the sun and currents of air for several days, and kept wet by sprinkling.

The method of construction of a small concrete tank is plainly shown in the accompanying illustration (Fig. 33). A tank of this size is suitable for ordinary purposes. See also Plate 4.

A Concrete Cistern. The mixture for a cistern should be made of 1 part cement and 3 parts sharp sand. The two should be mixed dry, and water sufficient to make a stiff mortar should be added. If the soil is of clay or of such a character that there will be no danger of a cave-in, it is not necessary to wall up with brick. The mortar should be applied about one-half inch thick, and should be followed immediately with a second coat one-fourth inch thick. Then give a skim coat made of equal parts of cement and sand. Keep the surface moist a week, and do not fill with water until 10 or 12 days.

Cisterns made in this manner are usually built egg-shaped, with the use of falsework for crowning the top, in which case the skim coat cannot be added. The walls in egg-shaped cisterns should be two inches thick at the bottom, three-quarters of an inch on the side, and three inches on the crown. If there is to be more than the ordinary pressure on the crown, it can be reinforced with wire netting. Local conditions must favor this method; it will not answer in all cases. A four-inch brick wall laid in cement mortar (one part cement and two parts sand), and plastered on the inside a half-inch thick with the same mortar, waterproofed, is in most cases satisfactory. But to make the cistern ab-

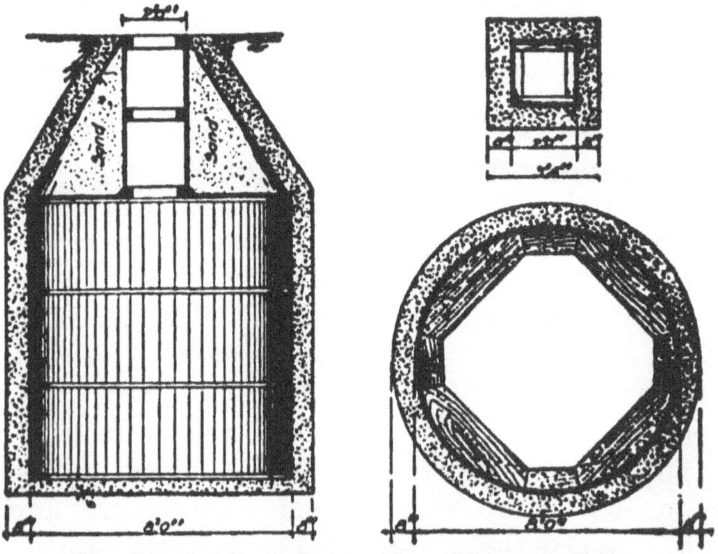

Fig. 84. Details of Construction of Concrete Cistern.

solutely water-tight, let the wall become hard, which requires six or eight days.

A square cistern is much easier to build, and in most cases answers the purpose as well as a round cistern.

Excavate to desired depth, and put in 6 inches concrete floor, 1 part Portland cement, 2 parts sand, and 4 parts broken stone. As soon as practicable, put up forms for 8-inch walls, and build the four walls simultaneously. If more than 8 feet square, walls should be reinforced with a woven wire fabric or steel rods.

Concrete Well Curbs. For cleanliness and sanitation, so essential around a well, concrete possesses advantages over brick or stone, especially if the surface down to the water line is made waterproof to keep out any possible seepage of surface water. A method that is recommended for the building of well curbs is here given that is applicable to almost all cases. The excavation having been carried down to water, the sides of the well made smooth and made ready for the concrete, a form ten inches less in diameter than the well excavation should be made of planks nailed securely and vertically to a frame. This form should be at least two feet in height, and may be higher, but a two-foot form will be handled more easily.

Operations should commence at the bottom of the well, where the form is first placed. The mixture for the concrete should consist of 1 part Portland cement, 3 parts coarse, sharp sand, and 5 parts gravel or broken stone. A waterproofing also should be used. Placing the form in the center so that a five-inch margin will remain around the circumference for the concrete, begin filling in, tamping every four inches. When the form is filled nearly to the top, allow the concrete to set, and raise the form carefully for the next section, being careful to have it at all times plumb. This operation is repeated with the filling in till the top of the well is reached. The concrete should be laid five or six inches above the surface of the ground to keep out all surface water. As the work of filling-in progresses, staging of planks can be built to provide working room. After the concrete is dried, these can be taken out.

HOLLOW-WALL CONCRETE SILOS

Throughout the northern section of the United States and in Canada, the introduction of the hollow wall—continuous dead air space—in concrete silo construction has met with general favor and approval. The advantage of this form of silo is apparent to those living in a country exposed to very low temperature for prolonged periods of time. With proper attention to the closing of doors, silage can be kept through the coldest winter without freezing. While freezing may not hurt the silage for food purposes, it certainly does not add anything of value to it. This, in connection with the trouble of thawing it out before feeding, makes the use of a silo that is proof against freezing, of a decided advantage. Wood silos with double walls have been built throughout the Northern States in an attempt to prevent this freezing; but while they have succeeded fairly well in this feature, have in a comparatively short time failed by the rotting of the wood. The silage juices get into the dead air space and rot out the wood in a remarkably short time. The only advantage claimed for the best double-wall wood silo—lined with cement plaster—has been its greater cheapness. With

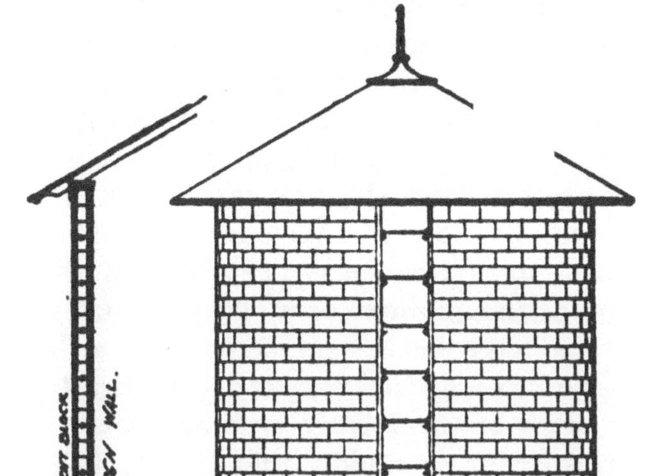

Fig. 34A. Concrete Block Silo.

the reduction in cost of the concrete silo, this advantage has now practically disappeared.

This reduction in the cost of the hollow-wall

type is due to the use of reinforcement and of specially adapted forms for moulding the green concrete.

Advantages of Concrete Silos. Fire is the farmer's greatest dread. When a fire starts from lightning or any other cause, the farm buildings usually burn down. For this reason, insurance rates are very high, and farmers find it a great tax to protect themselves by carrying insurance. The buildings, however, are not the most serious loss; fires most frequently occur during the latter part of the summer or in the early fall, after the crops have been harvested; and, although the buildings can be replaced, practically the year's work of the farmer is gone. Very often the fire spreads so rapidly that the stock is also lost.

A concrete silo cannot burn down; nor can the food stored in it be either injured or destroyed.

Insurance companies have recognized the indestructible qualities of concrete by making an insurance rate so low as to be within the means of every farmer.

The only objection that has ever been made to a concrete silo is its cost. The cost varies, owing to the price of materials of which the concrete is made. In many places concrete silos are cheaper than any other kind. Few farmers are without a gravel pit suitable to furnish both gravel and sand of a quality proper for making good concrete. Moreover, Portland cement can now be obtained at a reasonable cost. Under these conditions, a concrete silo is cheaper than any other kind.

The best is the cheapest, regardless of the original cost. A silo which never leaks, will not blow over, is always ready to be filled without first repairing, requires no repairs, cannot burn down, and is vermin-proof, is certainly the best and cheapest. A concrete silo is all of these.

Kinds of Concrete Silos. Three kinds of concrete silos have been in successful use for several years. These are known as monolithic solid-wall silos, monolithic hollow-wall silos, and concrete block silos. All three are good; and in choosing between them, the cost, which is fixed by local conditions, should be the deciding factor, unless the location of the farm is so far north that the extreme cold in winter makes a hollow monolithic wall or hollow block wall silo preferable to prevent the freezing of the silage. See Plate 5.

Size of the Silo. The size of the silo depends upon the number of cattle to be fed and on the number of days their feeding continues. It does not pay to build a silo for less than ten head; but, as someone very aptly put it, "Build a silo and get the ten head to keep."

The diameter, inside measurement, should never be more than one-half the height, and in practice it is not found advisable to make it over 20 feet.

For convenience, Table XVIII has been prepared, showing the size of silo required for feeding any number of cattle for a given time.

TABLE XVIII
Dimensions and Capacities of Silos

(Based on a feeding of 40 pounds of silage per cow per day.)

Number of Cows in Herd	FEED FOR 180 DAYS				FEED FOR 240 DAYS			
	Estimated Tonnage of Silage Consumed	Size of Silo		Corn Acreage Required at 15 Tons to Acre	Estimated Tonnage of Silage Consumed	Size of Silo		Corn Acreage Required at 15 Tons to Acre
		Diameter	Height			Diameter	Height	
	Tons	Feet	Feet	Acres	Tons	Feet	Feet	Acres
10	36	10	25	2⅖	48	10	31	3⅕
12	43	10	28	3	57	10	35	4
15	54	11	29	4	72	11	36	5
20	72	12	32	5	96	12	39	6⅖
25	90	13	33	6	120	13	40	8
30	108	14	34	7⅕	144	15	37	10
35	126	15	34	8⅖	168	16	38	11
40	144	16	35	10	192	17	39	13
45	162	16	37	11	216	18	39	14⅖
50	180	17	37	12	240	19	39	16
60	216	18	39	14⅖	288	20	40	19
70	252	19	40	17	336	..	..	..

This table gives the number of cows in herd, and tonnage of silage, for both 180 and 240 days of feeding of 40 pounds of silage per cow; also acreage of corn estimated to fill the silo, and the dimensions of the site itself. The diameters given are such that at least 2 inches in depth of silage will be taken off daily.

As stated above, the number of animals to be fed should determine the diameter of the silo, and the length of time silage is wanted should determine the height of the silo. The amount of silage to be fed per cow must be determined first. Decide whether each cow is to have 20, 30, 40, or 60 pounds per day. Then, having decided this point, make the diameter of the silo such that by feeding the cows so much per day the silage can be fed down at least 2 inches per day, as this will prevent moulding of silage.

Where large cows are kept, and it is expected to feed 40 or 60 pounds per cow daily, it frequently happens that it is desirable to cut down the silage ration. It is well to have the diameter of the silo small enough so that the farmer can cut down the ration one-third or even one-half, and still be able to feed down the silage 1 to 1½ inches daily.

In the dairying sections, many farmers consider this point so important that they are building two small silos instead of one large one, so that they can feed a light ration and still feed down the silage rapidly enough to prevent moulding.

Silo of Concrete Blocks. The method used for the construction of a silo of this character is as follows:

Place five eight-inch rods with turnbuckles every four or five feet, the ends being down in the hollow block sufficiently to hold while tightening and turnbuckle. The lap or tie rods used around the entire circle at the same course of blocks, are made of half-inch rods. The jamb at the opening is made of 2 by 12-inch wood, with a 1 by 4-inch board setting into the recess of the block to prevent slipping, and two stops on the opening side for a sliding plank. A casing board on the inside is required only when the blocks have no recess. It may prove cheaper to build a silo of 2 by 6-inch studding, 16-inch centers, with metal lath and stucco plaster on both outside and interior.

CONCRETE IN BARN AND STABLE CONSTRUCTION

Barn Foundations. These are laid in the same manner and of the same proportions in the mixture as house foundations, except that there is usually no cellar.

Barn and Stable Floors. The general rules of sidewalk construction, given under the head

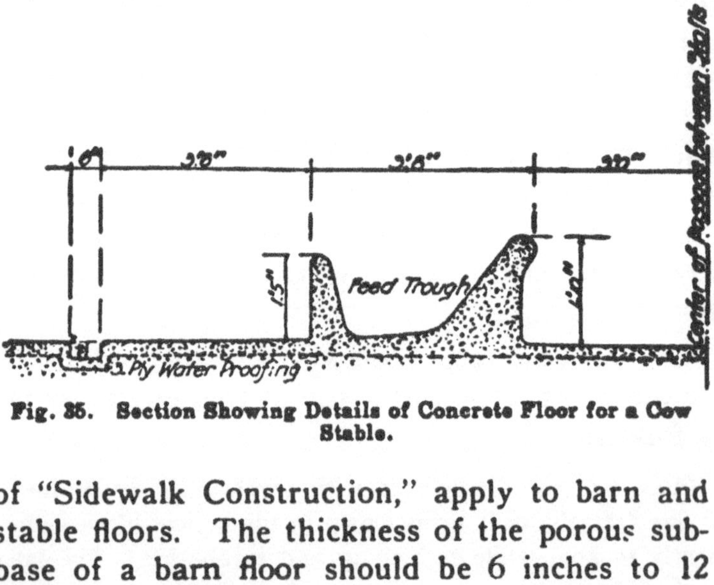

Fig. 85. Section Showing Details of Concrete Floor for a Cow Stable.

of "Sidewalk Construction," apply to barn and stable floors. The thickness of the porous subbase of a barn floor should be 6 inches to 12 inches, the base 3 inches to 5 inches, finishing with a surface of mortar, 1 part Portland cement and 1½ parts clean, coarse sand, 1 inch to 1½ inches thick. This may be roughed at time of laying and before it has set, or grooved in blocks about 6 inches square, to prevent the animals slipping. The surface should have sufficient slope to carry liquids to drains placed at convenient intervals. These drains may be either gutters or pipes laid under the floor, leading to a manure pit. If pipes are used, they should be laid in the sub-base, and the joints put together with cement mortar, care being taken to give the pipes enough slope to flush properly, and making them of straight lengths between openings so that they can be cleaned if necessary. The lids of the drain should be sunk about ¼ inch below the level of the floor, and should be loose, so that they can be removed conveniently.

Several years' experience in the use of concrete for barn floors and drains proves that manure will not injure well-made concrete, provided the concrete has thoroughly set and hardened before use.

Driveways are made by roughening or by dividing into 6-inch squares to give foothold.

The dairyman and agriculturist are more and more coming to recognize the concrete floor as the ideal for barn and stable. Excavation for a stable floor should be made below the frost line, and there should be a sub-foundation of at least six inches, and even more if possible, depending upon the weight and wear the floor is to have. A deposit of five inches of concrete should be made upon this sub-foundation, consisting of 1 part Portland cement, 3 parts sand, and 5 parts gravel or crushed stone, well mixed. The top coat should be two inches thick, 1 part cement and 2 of sand. The surface should be so sloped that the liquid manure and water of the stable will flow to some desired point for drainage away. The top should be grooved before it sets, to give the animals foothold and prevent their slipping. If the floor is to be of more than ordinary size, it should be laid in sections, and provided either with sand joints or the sections separated by pieces of tar paper.

Feeding Floors. The immense advantage of concrete feeding floors over the old method of placing fodder on the ground, is apparent to all who have given the subject any thought. Feeding floors should be built the same as sidewalks. The finishing coat is optional, although it has the advantage of being much easier to keep clean. Many farmers prefer an unfinished surface, on account of its giving cattle a firmer footing in slippery weather.

Piers and Posts. For this work the builder should excavate below the frost line, and build forms 2 feet square to a point within 6 inches of the surface of the ground. Fill with concrete consisting of 1 part cement, 2½ parts clean, sharp sand, and 5 parts gravel or crushed stone not more than one inch in size. The mixture must be tamped carefully. From the center of this foundation, build a hollow form one foot square and to the desired height, and fill with concrete of the same composition as the other. Before the form is filled, place four steel bars ¼ inch in diameter, vertically, so that they will be about two inches inside the corners; and around them, at intervals of one foot, wind loops of ⅛- or ¼-inch wire, tying them to the steel rods with finer wire. Every two feet a short piece of ½-or ⅓-inch wire may be tied to each of the ver-

Fig. 86. Use of Forms in Construction of Piers and Posts.

tical rods so as to project against the form and hold the steel in place. The concrete should be made soft and pliable so that it will flow, and, as it is poured into the top of the mould, work a long paddle, made like the oar of a rowboat, against the forms, to force the stones away from the surface and drive out bubbles of air which tend to adhere to the forms and form pockets. This method of construction makes an excellent foundation for a barn.

A Concrete Rain Barrel. Among the newer uses to which concrete has been placed is that of making rain barrels. A convenient size, easily made, is 36 inches high, 24 inches in diameter, with a shell ½ inch thick, the whole being treated with a waterproofing compound. No matter how long such a barrel is left empty in the hot sun, it will not spring a leak.

A Horse Block. In the construction of a horse block, the method does not differ materially from that employed in making a small tank. Build a box 24 inches long, 10 inches wide, and 8 inches deep, outside measure. Turn this bottom up on the floor or some other smooth surface; and around it build a box or form, without bottom, 36 inches long, 18 inches wide, and 12 inches deep, inside measure. Be sure that the smaller box is set at equal distances from both sides and ends of the larger box, and fill the form thus made with concrete, 1 part Portland cement, 3 parts clean, coarse sand, and 5 parts gravel or broken stone. Scrape with straightedge, and smooth with wooden float. Let it stand for at least 48 hours before removing outside form. Keep damp by sprinkling for three weeks., and do not attempt to move it before that time. If finished appearance is desired, as soon as the form is removed a coating one-eighth inch thick, made of one part Portland cement and one part clean sand may be plastered over the entire surface of the block, after picking with a stone axe and wetting thoroughly. See Plate 4.

Hot-Bed Frames. Excavate a trench to a depth below frost, and erect forms for a 4-inch wall. Fill with concrete mixture 1 part Portland cement, 3 parts clean, coarse sand, and 6 parts gravel or broken stone that will pass a half-inch sieve, to level of the ground. On top of these, build forms for a 3-inch wall to the height desired, and fill with concrete of the same proportions. These structures are so small in size that no reinforcement is necessary in the walls. On the upper edge of the walls, and around the interior, may be embedded strips for use when the glass frames are placed on top. Remove the forms in two or three days, and keep the walls damp for a couple of weeks.

Greenhouses. The concrete greenhouse offers the special advantage of being more easily heated than a wooden one. Greenhouses of this construction also keep out the cold air, and protect the growing plants against sudden changes of temperature.

The greenhouse foundation should be 10 inches wide and 16 inches deep. The mixture should be 1 part Portland cement, 3 parts sand, and 6 parts crushed stone or gravel. On this, a wall seven inches thick should be built. The mixture for this should be 1 part Portland cement, 2½ parts sand, and 5 parts cinders. The final or finishing dress for all the concrete should be a quarter-inch coat of cement mortar.

ROAD CULVERTS

It may be said of the concrete culvert that it will be found intact after a flood or freshet, while brick or stone culverts in the same neighborhood will be washed away. It is true, too, that they are cheaper than culverts made of any other material, besides being more durable. It need only be mentioned that the only time to build a culvert is in a dry season when there will be no difficulty with water. When this is not practicable, the water should be diverted, if possible.

In this work as in all other operations where concrete is to be placed on the ground, excavation must be carried below the frost line. Trenches for the foundation should be dug on each side of the bed of the stream. The concrete should be of 1 part Portland cement, 3 parts sharp, clean sand, and 6 parts gravel or crushed stone. Build an apron with this concrete across the bed of the stream betwen the two foundations, and with its level equal to the bed of the stream. For a culvert of ordinary length this apron should be eight inches thick. Place the semicircle arches of the size required, and brace them well. Two-inch plank should be used

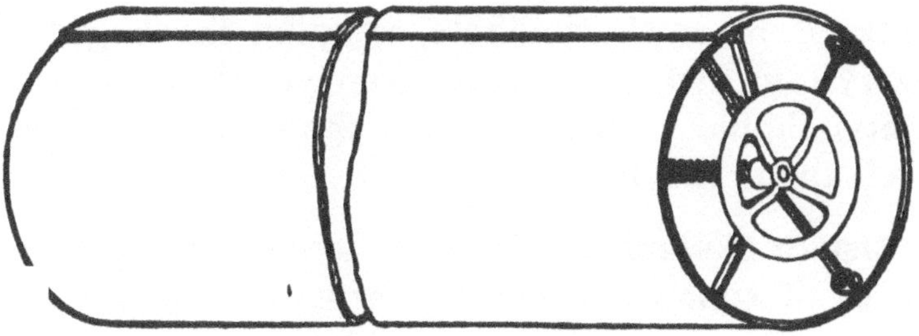

Fig. 37. Collapsible Form.

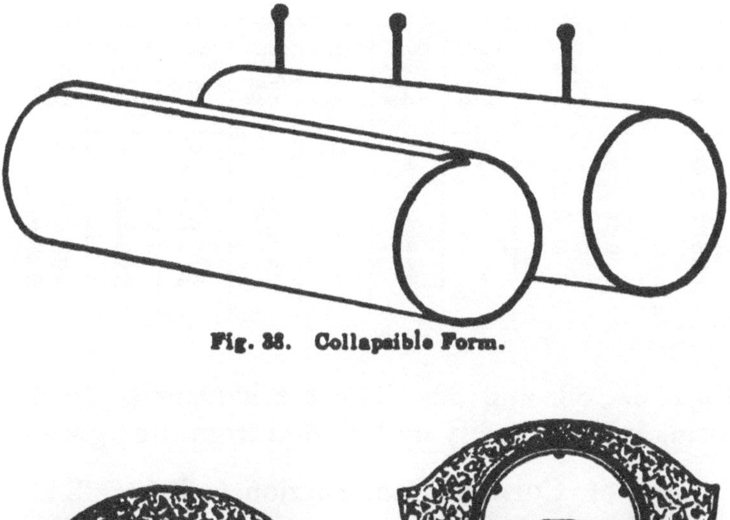

Fig. 38. Collapsible Form.

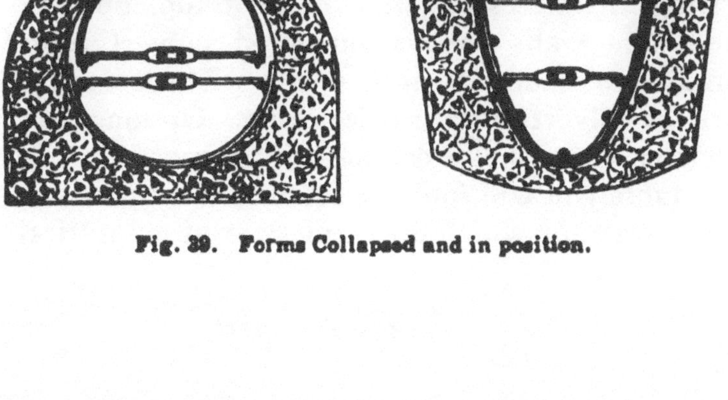

Fig. 39. Forms Collapsed and in position.

for the forms. After the forms have been well greased, fill in the concrete, tamping thoroughly every four inches. A reinforcement of expanded metal should be placed inside the surface of the arch about two inches.

Use of Collapsible Forms. Some of the many types of collapsible culvert forms are shown in

TABLE XIX

Data for Estimating on Culvert Construction.

Diameter of Culvert Inches	Waterway Squared Inches	Thickness Bottom Inches	Thickness Sides Inches	Thickness Top Inches	Cubic Feet Sand Required	Pounds Cement Required	Width to Dig Ditch Inches	Complete Cost
[illegible]	6x 8.37	3	[illegible]	4½	30	465	16	$ 5.40
11	10x 9.50	3	4	5	36	558	19	6.84
13	12x11.60	3	5	6	48	744	23	8.64
15	12x14.75	4	[illegible]	6	59	915	25	10.63
19	16x17.72	4½	[illegible]	[illegible]	85	1317	30	15.39
22	19x20	[illegible]	5½	6½	90	1395	33	16.30
27	24x23.75	5	6	7	124	1922	39	22.33
36	24x41		6	7	165	2557	48	29.70
48	36x57.50	6	7	8	255	3952	68	49.90

Figs. 37, 38, and 39. The mechanism of these forms will be readily understood from the figures.

Cost of Culvert Construction. Table XIX gives figures from which to make estimates, giving the necessary thickness of top, bottom, and side walls for various-sized culverts, the amount of material needed, etc. These figures are for culverts eighteen feet long. On longer or shorter ones, make estimate in proportion. In the table you will find the squared dimensions of waterway for which different sizes of cylindrical moulds will be the equivalent; also the thickness of the sides, bottom and top walls, the width to dig the trench, and the amount of sand or gravel. and of cement to be used.

The costs of completed culverts as here estimated are figured on what are considered average conditions, with cement at 65 cents per hundred pounds, and gravel at $1 per yard. The cost of this kind of concrete work is close to 18 cents per cubic foot of concrete used; and, as cement in the proportions mentioned does not increase the bulk of gravel, to estimate the cost of any sized culvert completed, multiply the estimated amount of gravel in cubic feet by 18—this will give you the cost complete. Multiply by 11 to get the cost of cement; by 7 for cost of material on the ground and placed.

The figures in the table are made on a basis of 18-foot culverts, with coping to extend two feet from waterway through culvert, and concrete in proportion of 1 part cement and 6 parts gravel or sand.

To **Keep the Road Open.** For convenience in keeping the road open for traffic, and the saving in material for forms, we suggest making only nine feet of the culvert at a time. Should this suggestion be accepted, proceed as follows:

Make three semicircular forms the size required, out of 1½-inch stuff, and set them in place three feet apart. Fasten joist 2 inches by 4 inches by 9 feet on them. This is called **lagging.** Set the form thus made on large wedges supported by top of form. Grease forms well, and fill with concrete of a rather wet consistency, and tamp thoroughly every 6 inches, taking care not to disturb the form. Let stand until thoroughly dry, about 28 days; and then knock out the wedges, lowering the semicircular form, which will be easy to remove.

Should the culvert be made all at one time, enough semicircular forms should be constructed to support the lagging at least every 3 feet.

Reinforce the concrete with expanded metal, placing it so that it is 2½ inches in from the under side of the arch, and extending down through the walls. All concrete should be mixed 1 part Portland cement, 3 parts sand, 6 parts broken stone. Should wing-walls be required, they should be built at the same time as the foundation, should go to the same depth, and be reinforced, the reinforcing connecting with that in the walls. The width of these walls should be left to the judgment of the man in charge of the work.

CEMENT DRAIN-TILE AND SEWER PIPE

Cement tile have abundantly demonstrated their efficiency and durability for drainage purposes. Their permeability is a feature that can be controlled by the manufacturer. Tile can be made porous so as to give effective drainage throughout their length; or they can be made water-tight so as to admit water only at the joints, like ordinary clay tile. If properly made of well-selected material, they grow stronger with age. In many cities, concrete sewers are replacing brick sewers on account of the more satisfactory wearing qualities of the former. When properly cured, cement tile will withstand alternate freezing and thawing without injury, even when partially immersed in water. In the construction of a county ditch near Emmetsburg, Iowa, nearly a mile of cement tile lay stretched out through a swamp for eighteen months, or through two winters, before placing in the work. Some of these tile were entirely submerged, others partially, and some were entirely out of water. They were subjected to freezing and thawing, and yet remained in perfect condition.

On account of their thin walls, cement pipe and tile are not so easily made as cement blocks. Tile machines, however, are now made with all the skill and ingenuity embodied in block machines, and with perfectly efficient adaptation to their purpose.

Tile are made with different types of joints—ordinary butt joints, tongued-and-grooved joints, and bell-end joints, the last named being made either "hell up" or "bell down."

Rudolph Hering, a New York engineer, who has investigated the subject, claims the following advantages for concrete sewer pipe as compared with vitrified clay pipe:

A sectional form can be given them which is more conducive to stability and efficiency than the round form customary in clay pipes.

As vitrified pipe warp in burning, the section is not finished truly circular, and slight projections are formed at every joint when the pipes are laid to form a sewer.

As cement pipes have a truer sectional shape than vitrified pipes, they can be given a slanting butt joint, as is customary in Europe, instead of the more costly bell and spigot joint common for vitrified pipe, which are made in imitation of cast-iron pipe used under high pressures.

Concrete pipes are tougher and less brittle than vitrified pipes.

Concrete pipes, if well made of proper materials, have a strength to resist compressive, tensile, and bursting strains which is amply sufficient for all purposes for a sewer in a large city. If the materials are carefully selected, the concrete pipe should be as permanent as the vitrified pipe. Concrete work in the sewers of Paris several hundred years old is as sound to-day as when it was laid.

For the manufacture of concrete sewer pipes, a number of machines are on the market; and when the aggregate used is carefully selected and cleaned, the mixture being of 1 part Portland cement, 2½ of sharp, coarse sand, and 4 parts crushed stone not over half an inch in size, a good pipe should be turned out.

Circular forms of steel should be used, and the tamping should be done in the same manner as for concrete blocks.

Concrete Chicken-Houses

It is easier to keep a chicken-house clean when it is made of cement than it is to care for such a house built of any other kind of material. Besides, the concrete chicken-house is rat-proof.

Excavate a trench 12 inches wide, to a depth below frost, and fill with concrete 1 part Portland cement, 3 parts clean, coarse sand, and 6 parts cinders. On this foundation, build a solid wall 5 inches thick, 1 part Portland cement, 2½ parts clean, coarse sand, and 5 parts cinders; or, if cinders are not obtainable, a hollow wall should be built 12 inches thick, consisting of two 3-inch walls and a 6-inch air space. The roof may be made of wood or of concrete. If the house is not more than 8 feet wide, a roof with slope in one direction may be made of a 4-inch concrete slab reinforced with steel rods or heavy wire mesh. For a shorter span, a less thickness may be adopted.

Hens' nests are also made of concrete. They are vermin-proof, and are adapted to maintaining the desired evenness of temperature. They can be washed out, or kept in perfect sanitary condition by filling with straw or other combustible material, and burning out.

HOG-PENS OF CONCRETE

Usually on the farm, the greatest difficulty in the matter of cleanliness about the buildings is encountered in connection with the care of the hog-pens. By the use of concrete, whose surface lends itself readily to flushing with water, much of this trouble is avoided.

The size and shape of the pen having been decided upon, a trench for the foundation should be excavated a foot wide and below the frost line. In this, a foundation consisting of 1 part Portland cement, 3 parts sand, and 6 parts gravel or crushed stone, should be laid. Four parts of gravel or crushed stone is used in the mixture for the wall, instead of six parts as in the foundation. The floor of the pen is made in the same manner as followed in the laying of concrete walks.

The hog-house may be built in one corner of the pen, and its walls should be four inches thick, with a reinforced one-slant roof of 2½-inch concrete reinforced with netting and with rods of ½-inch thickness placed 10 inches apart, if the house is not more than 10 by 12 feet.

A trough can be made by the use of two forms, one of a long box shape, and the other shaped

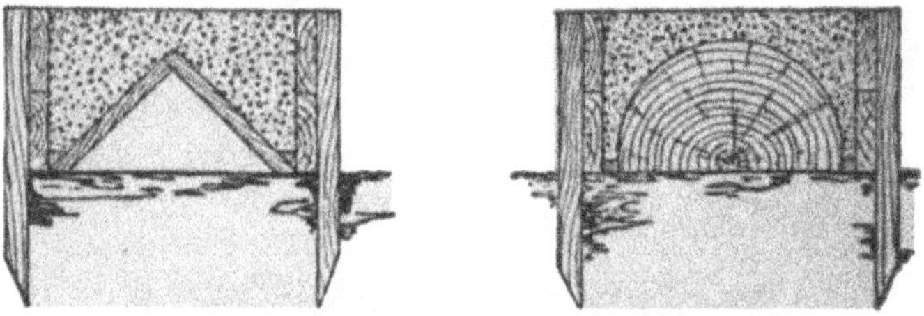

Fig. 40. Forms for Hog Troughs.

like a V for the trough cast. First provide a smooth surface or platform to lay the forms on. Place first the V form in the position of a pyramid. Over it set the box, being careful that the adjustment is arranged so the V form is in the middle of the box. Fill with a mixture consisting of 1 part Portland cement, 3 parts sand, and 5 parts gravel or crushed stone. Do not disturb the forms for three or four weeks.

Concrete Ice-Houses

The modern equipment of a rural home, especially if it is located where a supply of ice is available for harvest in the winter, is not complete without an ice-house. And for this purpose no material lends itself better than does concrete. To be sure that the heat will be kept out in the summer, it is well to provide for a wall at least sixteen inches in thickness. A wall of this character will protect the contents of the ice-house amply. A hollow wall, also, is advantageous.

For an ice-house of ordinary size, sufficient to hold a supply for a family of five or six, an excavation one foot below frost will do. This surface should be cleaned off and leveled, and upon it a layer of crushed stone or broken bricks should be placed, ramming the material thoroughly. This will afford opportunity for drainage.

If the wall is to be sixteen inches thick, the forms should be set up allowing for a space of that width.

The foundation, on the sub-foundation that has been previously wetted, should be composed of a mixture of 1 part Portland cement, 3 parts clean, sharp sand, and 6 parts broken stone. The foundation ought to be four feet deep. Better satisfaction will be secured if provision is made for an air-space between the walls. Two 6-inch walls 4 inches apart, and bound together with rods, will be a good arrangement. Separate forms for each must be constructed. Walls of this width will require no other reinforcement than the binding rods, provided the house is not to be high. One part of Portland cement, two of sand, and four of crushed stone will be the proportions of the mixture for the walls. The walls should be built in sections about two feet high at a time. Place the rods of half-inch iron with strong heads imbedded about two inches in the inner surface of each wall, and about a foot apart. This will help the wall stand the lateral pressure of any pile of ice within that may rest against it.

The roof, reinforced with ¾-inch iron rods a foot apart, is the next step. A form is constructed, of the desired angle. Upon this, about three inches of sand is placed and smoothed off carefully. Lay the rods so that they will rest 1½ inches above the sand, and put on a coat of three inches of concrete. The forms should not be touched for two weeks. Then the sand can be let out from the interior. All the openings between the walls and the roof must be closed up.

Small Storage Buildings. These should either be built of hollow concrete blocks, or, if monolithic in form, there should be an air-space in the walls. The air-space is necessary to retard dampness and keep out the frost.

Root Cellars. Cellars of this character usually are built half below and half above the surface of the ground. When properly made, they ought to be proof against all inroads of frost. Excavation should be carried down to a point sixteen inches below the desired level of the floor. The foundation should be twelve inches wide; and the mixture for it should be one part Portland cement, three parts coarse sand, and six parts gravel or crushed stone. Forms should be made for the foundation. After it is laid, the forms should be removed; and a porous material, either broken bricks or cinders, should be filled in over the floor space to a depth of twelve inches. This should be well tamped. The floor, of a total thickness of four inches, sbould consist of three inches of concrete and one inch of cement mortar. The concrete should be one part cement, three parts sand, and five parts gravel or crushed stone.

The wall of the cellar should be eight inches thick, started from the center of the 12-inch foundation. The mixture for the wall may be 1 part Portland cement, 2½ parts sand, and 5 parts crushed stone, gravel, or cinders. Build up the end walls so as to form a point at the middle, and high enough to give the roof a sufficient pitch to shed the rain. Near the top at each end, it should be remembered to provide openings for windows; and the sash should be fitted and plastered in after the concrete has set and forms have been removed. Bins should be built of a size to suit convenience, with walls four inches thick and reinforced with one-quarter-inch rods placed twelve inches apart horizontally and vertically to give the bin walls strength to withstand the lateral pressure when they are filled with vegetables.

If a concrete roof is desired, forms should be erected, and a roof two and one-half inches thick built. On the top, and before the concrete is dry, a quarter-inch layer of mortar, consisting of one part Portland cement and one part sand,

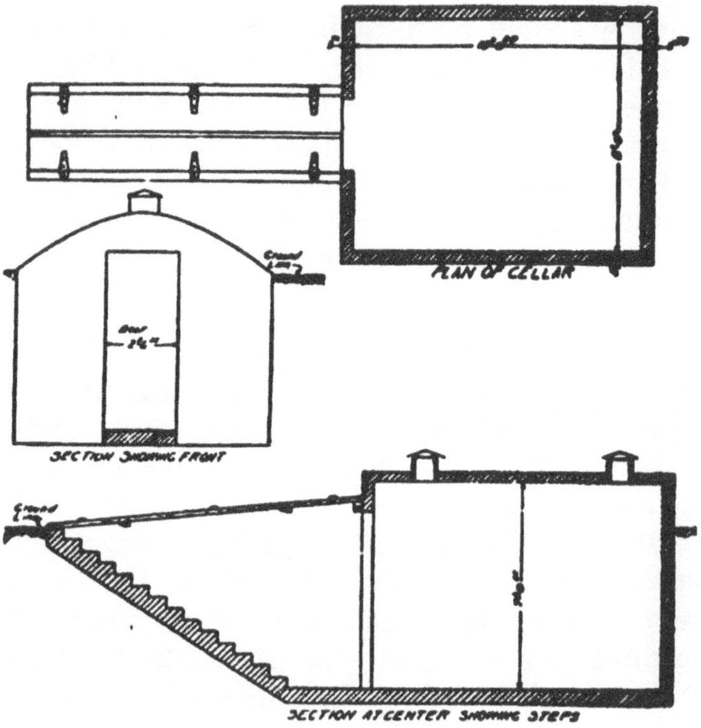

Fig. 41. A Cyclone Cellar.

should be placed and well troweled. The forms should not be removed before three weeks. The roof should be reinforced with woven wire fabric, and so should the steps. If the roof is sufficiently long to require supports, an eight-inch pillar in the center may be erected, reinforced with one-half-inch rods two inches apart and one inch from the surface.

Mushroom Cellar. The method of constructing a cellar of this character is the same as for a root cellar, with the exception that no floor will be required, and there should be little light.

Cyclone Cellar of Concrete. Next to fire, a farmer on the western plains fears a cyclone more than anything else on earth. The only sure salvation is to get below the surface of the ground, and practically every farm has a cyclone cellar. It is becoming the practice to construct these places of refuge of concrete. In using concrete as a building material, there is no danger of the roof blowing off, or of the walls rotting out in a few years and having to be renewed. Fig. 41 shows the design for such a structure.

CONCRETE FENCE POSTS

There is a constantly increasing demand for some kind of fence-post which is not subject to decay. The life of wooden posts is very limited, and the scarcity of suitable timber in many localities has made it imperative to find a substitute. A fence-post, to prove thoroughly satisfactory, must fulfill three conditions:

(1) It must be obtainable at a reasonable cost;

(2) It must possess sufficient strength to meet the demands of general farm use;

(3) It must not be subject to decay, and must be able to withstand successfully the effects of water, frost, and fire.

Although iron posts of various designs are frequently used for ornamental purposes, their adoption for general farm use is prohibited by their execssive cost. Then, too, iron posts exposed to the weather are subject to corrosion, to prevent which necessitates repainting from time to time; and this item will entail considerable expense in cases where a large number of posts are to be used.

Cost and Desirability. At the present time the material which seems most ready to meet these requirements is reinforced concrete. The idea of constructing fence-posts of concrete reinforced with iron or steel is by no means a new one; but on the contrary such posts have been experimented with for years, and a great number of patents have been issued covering many of the possible forms of reinforcement. It is frequently stated that a reinforced concrete post can be made and put into the ground for the same price as a wooden post. Of course this will depend in any locality upon the relative value of wood and the various materials which go to make up the concrete post; but in the great majority of cases, wood will prove the cheaper material in regard to first cost. On the other hand, a concrete post will last indefinitely, its strength increasing with age, whereas the wooden post must be replaced at short intervals, probably making it more expensive in the long run.

In regard to strength, it must be borne in mind that it is not possible to make concrete posts as strong as wooden posts of the same size; but since wooden posts, as a rule, are many times stronger than necessary, this difference in strength should not condemn the use of reinforced concrete for this purpose. Moreover, in many cases strength is of little importance, the fence being used only as a dividing line; and in such cases small concrete posts provide ample strength, and present a very uniform and neat appearance.

In any case, to enable concrete posts to withstand the loads they are called upon to carry, sufficient strength may be secured by means of reinforcement; and where great strength is required, this may be obtained by using a larger post with a greater proportion of metal and well braced as is usual in such cases. A post constructed of concrete reinforced with steel will last indefinitely and require no attention in the way of repairs.

Reinforcement. No form of wooden reinforcement, either on the surface or within the post, can be recommended. If on the surface, the wood will decay; and if a wooden core is used, it will in all probability swell by the absorption of moisture, and crack the post. The use of galvanized wire is sometimes advocated; but if the post is properly constructed and a good concrete used, this precaution against rust will be unnecessary, since it has been fully demonstrated by repeated tests that concrete protects steel perfectly against rust. If plain, smooth wire, or rods are used for reinforcement, they should be bent over at the ends or looped to prevent slipping in the concrete. Twisted fence wire may be obtained at a reasonable cost and is very well suited for this purpose. Barbed wire has been proposed, and is sometimes used, although the barbs make it extremely difficult to handle.

For the sake of economy the smallest amount of metal consistent with the desired strength must be used; and this requirement makes it necessary to place the reinforcement near the surface, where its strength is utilized to the greatest advantage, with only enough concrete on the outside to form a protective covering. A reinforcing member in each corner of the post is probably the most efficient arrangement.

Mixture for Concrete Posts. The concrete should be mixed with Portland cement in about the proportion 1:2½:5, broken stone or gravel under one-half inch being used.

How to Preserve Gate-Posts. A sagging gate-post, rotted at the ground line, gives the whole place an unkempt and rough look; and yet gate-posts do rot, and have a habit of getting out of line. The remedy is a very simple one, and very inexpensive as well. First brace the post in such a way as to prevent its falling. Then excavate around it, to a depth below the frost line. Pull the post into the proper position, and renail your braces. Fill the hole with concrete, to a point six inches above the ground, and your post will remain constantly in position. It will also last for years. When it has to be renewed, the old post can easily be pulled out, and a new one slipped into the hole in the concrete.

The materials needed for these repairs are: 1 bag of Portland cement; 3 cubic feet of gritty,

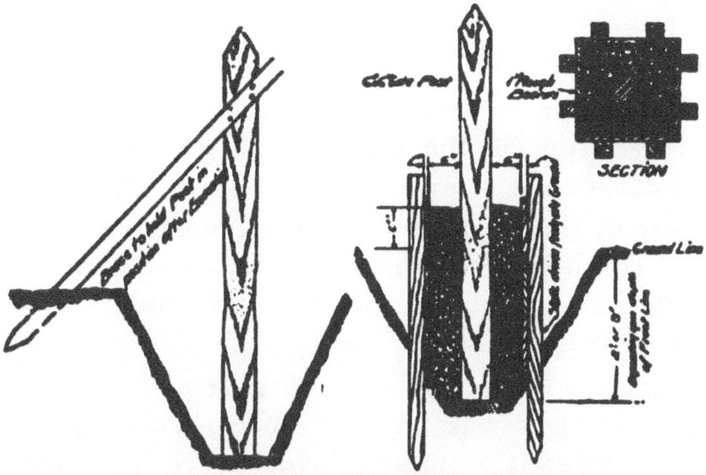

Fig. 43. Method of Preserving a Gate Post.

clean sand; 6 cubic feet of gravel, of a small size.

After excavating around the post, and bracing in position, drive stakes, and place against them rough boards, allowing the boards to come six inches above the ground line. This makes a box or form around the post, into which the concrete is placed.

Shovel the concrete directly inside the box, and tamp with a 3-inch by 4-inch piece of lumber. When filled, take a trowel and smooth off the top, leaving the whole mass slightly higher against the sides of the post than at the edges.

After two days, remove the braces and the forms, and fill with earth around the concrete, up to the ground level.

It will take one man about half a day to do all the work. The materials, except the cement, will cost nothing.

MISCELLANEOUS USES OF CONCRETE

One of the many novel uses to which concrete has been put is the construction of large **coal pockets** for the storage of anthracite coal.

One of these pockets has been constructed for the Lehigh & Wilkesbarre Coal Co., at Charlestown, Mass., and is capable of holding 10,000 tons of coal.

Concrete Chimneys. The first reinforced concrete chimney was built in 1898 for the Pacific Coast Borax Company, Bayonne, N. J. Since that time about 400 stacks have been completed. There are examples in nearly every State of the Union and in Canada. These stacks range in height above ground from 50 to 352½ feet, with inside diameter ranging from 4 to 18 feet, the majority of them being 150 to 200 feet high and 5 to 6 feet inside diameter.

A Concrete Lighthouse. That reinforced concrete can successfully resist a violent earth-shock, gales of hurricane-like force, and the combined fury of sea and tides, has been most satisfactorily demonstrated in the case of the little "Mile Rock" light station at the entrance to San Francisco Bay, California.

Penstocks. A comparatively new and relatively untried opportunity for the use of reinforced concrete is presented in the construction of penstocks for the conveying of water, frequently under considerable head.

Sewers and Conduits. Concrete has been extensively used in the construction of conduits

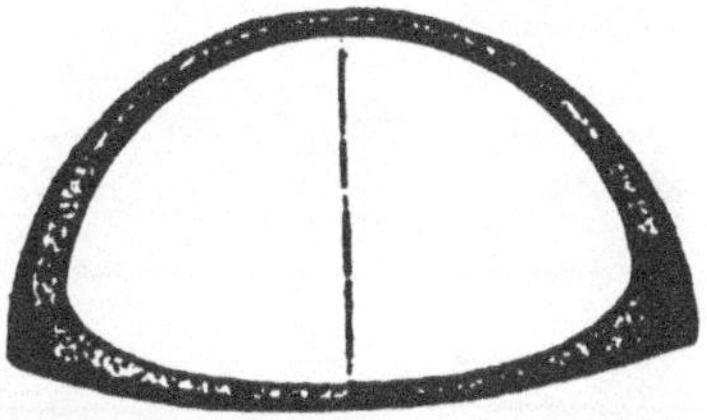

Fig. 44. Conduit Reinforced with Sheet Fabric.

and sewers, and its use is becoming more general. No one questions its suitability for the construction of aqueducts, conduits, and storm sewers; but doubt is sometimes expressed as to the ability of concrete sewers to withstand the possible action of acid sewage.

Concrete sewers are usually built in place; pressure pipes and small sewers—say those under three feet in diameter—are often laid in sections which have been moulded in advance, like drain-tile. This latter method of construction is economical on account of the thin sections of concrete that may be used, accurate forms and shop methods permitting the adoption of thinner sections than could be constructed in place.

Concrete Railroad Ties. The problem of the future supply of ties is one of the most serious that confronts the railroads. It should be remembered, too, that the widespread building of interurban trolley lines throughout the country has also tended to drain the visible sources of supply.

But there is a solution of the problem. Cement has been found by actual tests to make the best

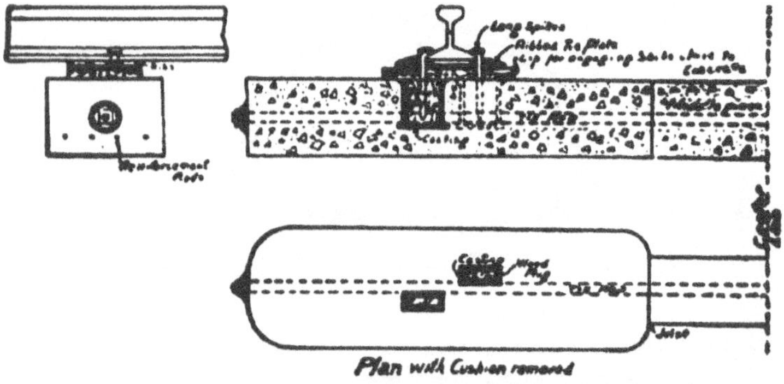

Fig. 45. One Type of Concrete Railroad Tie.

possible tie. It makes a tie, too, that will not wear out and never have to be renewed. Experiments on a number of roads have proven the actual serviceability of cement for this purpose.

Telegraph and Telephone Poles. These are now made of concrete and are found not only to out-last the ordinary wooden poles, but to have greater strength. Carefully calculated accounts of all expenditures for labor and material in the construction, show that under average conditions the first cost is slightly more than that of cedar poles. With concrete poles, however, the renewal cost is eliminated.

Concrete Docks and Piers. Concrete is coming into more favor year by year for the construction of docks and piers. With their cost but little more than that of timber, and their length of service beyond estimate, it is no wonder that keen business men realize their value.

Burial Vaults. Even these are now made of reinforced concrete. It was found that adjustable reinforcing forms for burial vaults was not a success; so now it is common to make them in two sizes—adult and child size. The whole body of the vault is made of expanded sheet steel centered in the highest grade of Portland cement concrete. The vault is waterproofed thoroughly after it is finished, white enameled in some instances on the inside with a waterproof paint and finished with aluminum waterproof paint on the outside.

Reinforced Concrete

When steel, or other material which is capable of resisting a tension, is embedded in concrete, thereby giving the mass a greater strength and elasticity, the process is known as **reinforcing**; and concrete thus treated is known as **reinforced concrete.**

Steel has about the same strength in tension when used as a beam, as it has in compression when used as a column or post. The same thing is approximately true of wood and some other materials of construction. In concrete, however, the conditions are quite different, the compressive resistance of concrete being about ten times its tensile resistance.

In a concrete beam, the upper portion of the beam is in **compression**, and the lower part is in **tension.** The line where the internal stresses of the beam section change from compression to tension is called the **neutral axis.**

The forces must balance on each side of the neutral axis. A plain concrete beam, being so much stronger in compression than in tension, will have its neutral axis located very low.

Steel is so much stronger in tension than concrete that a very small steel rod or bar placed in the bottom of a concrete beam will raise the neutral axis and balance the compressive forces exerted above.

Plain concrete, when used in the form of pillars and posts, is capable of carrying heavy direct loads through its great compressive strength. But when it is subjected to a direct pull—that is, to tensile strains—it is weak. For example, if a plain concrete beam is subjected to a load, it will break apart at the bottom just as a piece of chalk would under like conditions, being unable to resist the tension in the lower portion of the beam. In order to overcome this, reinforcing steel is used to give proper tensile strength and elasticity. The concrete in the top of the beam takes care of the compression. A properly reinforced concrete beam, therefore, has the strength of stone in resisting compression, united with the tension-resisting power of steel.

When a beam is loaded and supported at the two ends, it will have a tendency to deflect or bend. To illustrate, assume that a beam is made up of a series of flat plates—or, in other words, like a pad of paper or a book—the difference being that in the pad of paper the leaves are not in any way connected with one another, whereas in a beam the adhesion or sticking together of the various particles of the material ties the imaginary plates together. Now, when the supposed beam starts to deflect, one of two things will happen: either the various plates separate, as when a book or pad of paper is bent, and, in separating, slide by one another; or, if the plates are held together and sliding is prevented, the particles in the upper plates compress, and those in the lower plates elongate or stretch out.

It is thus seen that in addition to the compression and tensile stresses in the top and bottom of the beam, there are internal stresses of

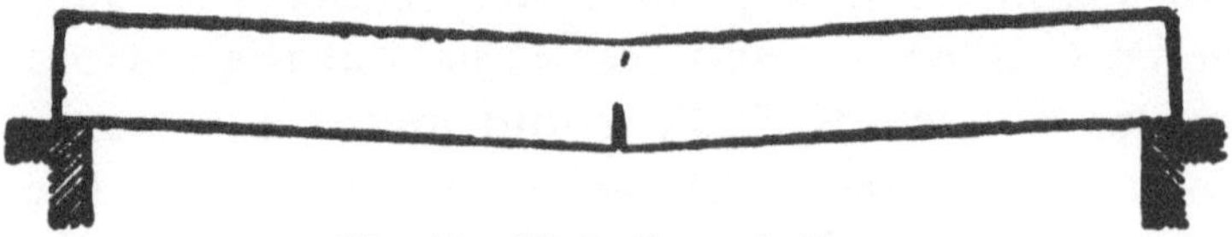

Fig. 46. Plain Concrete Beam.

equal importance, against which the concrete must also be properly reinforced. These may be **tensile** or they may be **shearing** forces.

Fig. 46 shows a plain concrete beam, supported freely at the ends, which has failed upon the application of a small load applied near its center.

Fig. 47 shows a similar beam having horizontal reinforcing rods located near the bottom sur-

Fig. 47. Beam with Horizontal Reinforcing Rods.

face of the beam. The method of failure under a medium load, in this case, was said to be due to the ends of the reinforcing rods **slipping** in the concrete. The diagonal or slanting cracks are partially due to horizontal shear set up by the bending of the beam. These are sometimes spoken of as due to **diagonal tension**.

A means of fortifying against horizontal shear is by the use of **stirrups.** Bands or rods of steel or iron are bent in the shape of a U, and either placed loosely around, fastened rigidly to, or made as a part of the reinforcing rods.

Fig. 48 shows the method of failure of the same type of beam as previously shown, but having loose stirrups surrounding the horizontal reinforcement bars and embedded vertically in the concrete. This beam failed when tested to destruction, by the slipping of the horizontal rods. The figure shows the shearing of the concrete

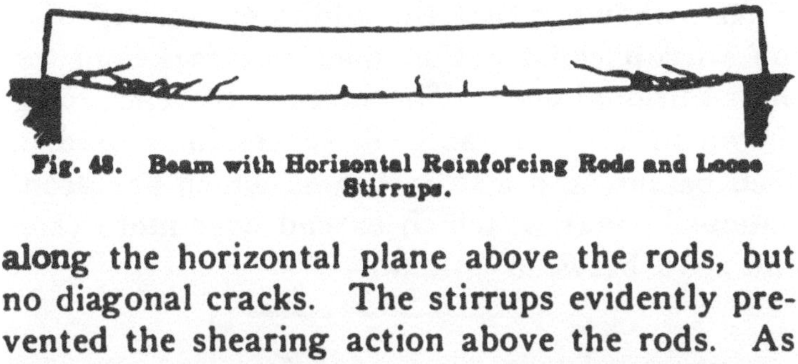

Fig. 48. Beam with Horizontal Reinforcing Rods and Loose Stirrups.

along the horizontal plane above the rods, but no diagonal cracks. The stirrups evidently prevented the shearing action above the rods. As a means of preventing such a method of failure, some companies have either rigidly fixed the stirrups to their reinforcing bars or formed them as a part of same. The result of such a construction seems to throw the greater part of the body stresses of the beam onto the horizontal bars for support. Some authorities consider this a weakness in the construction.

Fig. 49 shows the failure of the same type of beam as previously shown, but provided with reinforcing bars with fixed stirrups. From the cracks shown near the bottom of beam, this this seems to be a well-balanced reinforcement, with the main stress occurring at the center of the horizontal rods.

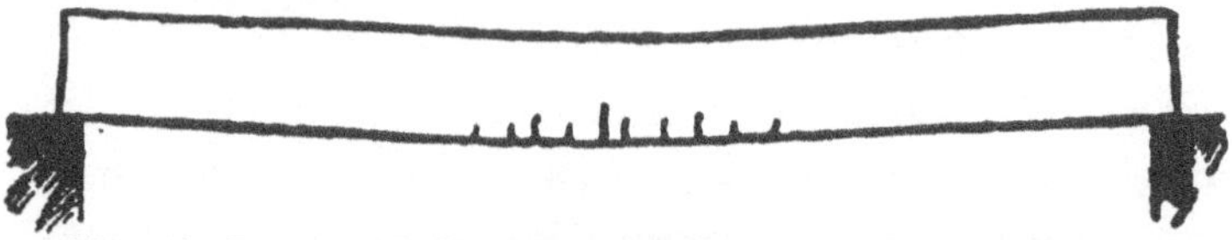

Fig. 49. Beam with Horizontal Reinforcing Rods and Fixed Stirrups.

While these different forms of reinforcement seem to be favored by many, they are criticised by some investigators who claim to have records of tests showing that the additional strength of the stirrup construction does not make up for the additional cost. The method of reinforcing shown in Fig. 50 has been successfully used in deep beams, and also in beams which are "continuous"—that is, which extend over more than one span between columns.

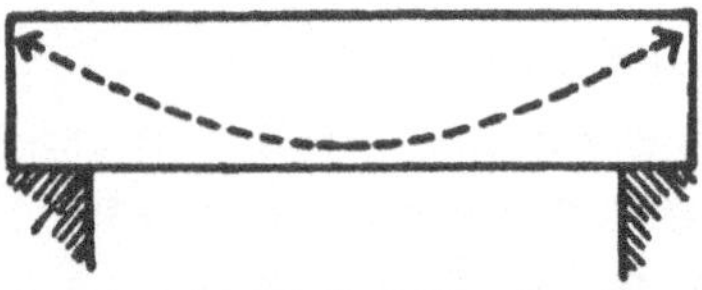

Fig. 50. Anchored End Bar Construction.

In Fig. 50, the method of fastening the ends of the reinforcement is a point to be noticed. In case of a continuous beam, the reinforcement would simply extend upwards toward the end of the column, and over into the next span. It is claimed by many authors that the ends of all rods for reinforcement should be either bent over and embedded in the concrete, or fitted with some kind of expanded end, to prevent the ends slipping when the beam is bent.

The most important principle in placing reinforcement in concrete beams is to place the steel so that it will relieve the concrete from all tensile stresses if possible, and thus aid in developing the high compressive strength of the material.

Every ounce of tension in the steel is only effective as it is transferred to the concrete. In the case of a plain beam with free ends, there is no stress in the steel at the ends, while the maximum tension is usually at or near the center of the beam. The entire amount of this tension must be gradually transferred from the steel to the concrete.

While the adhesion or sticking of the concrete to the steel is relied on to permit the transfer of this stress from one material to the other in much of the reinforced concrete work now being done, it is realized that this adhesion is not always permanent. Failures of floors have already occurred, due to loss of the adhesion, after they have successfully supported heavy loads for many years, the adhesion being greatly reduced with age and under certain unfavorable conditions, such as continued soaking of the concrete in water, long-continued vibration, etc.

Experience has demonstrated that beams may fail in other ways than by the pulling in two of the reinforcing steel, as. for example, by shearing across a vertical plane, by tension along a diagonal plane, or by slipping of the rods through the concrete.

One of the first uses of reinforced concrete in building construction was in the house erected by W. E. Ward in 1872, at Port Chester, N. Y. However, some twenty years earlier than this, in France, the first combinations of iron imbedded in concrete were made in a small way. But it was not until the very end of the last century—in fact, after 1895—that reinforced concrete came into extensive commercial use in the construction of buildings. Previously to this, the plain type of construction had attained a wide use in foundations, and at this time its development was beginning for such structures as dams, sewers, and subways.

Briefly, reinforced concrete such as is used for construction of industrial buildings, bridges, retaining walls, etc., consists of Portland cement, sand, and gravel or broken stone, mixed with water to a consistency that will just flow sluggishly, and in which steel rods are embedded so as to produce an artificial stone having many of the characteristics of steel.

In the early stages of reinforced concrete, and even up to the present time, many patents of a more or less fundamental character have been granted. These have taken the line of special forms of reinforcing metal as well as methods of design. Some of the principal styles of reinforcement are illustrated below, under "Reinforcing Systems."

ADVANTAGES OF REINFORCED CONCRETE

Reinforced concrete possesses many advantages that other building materials do not, and these have led to its rapid growth as a standard for many types of structures. The chief advantages that reinforced concrete has are as follows:

Its moderate cost of construction—less than that of steel, and only slightly greater than that of wood.

Its remarkable fire-resisting qualities, that have been shown in many instances.

Its strength and its capacity for resisting shock, due to the monolithic or one-piece nature of the structure.

Its freedom from rotting, to which wood structures are subject in course of time.

Protection afforded by the concrete to the reinforcing steel, which would corrode rapidly if left exposed.

Its capacity for resisting the action of many chemical compounds that would soon destroy structures made of wood or steel.

Reinforced concrete's low first cost has led to its use in preference to masonry and steel construction. While wood structures are cheaper than concrete, the latter are to be preferred on account of their superior fireproof qualities and their freedom from decay caused by rot and the attacks of vermin and insects. Fire insurance rates for reinforced concrete buildings are only about half those for wooden buildings of the type known as "slow-burning mill construction." The cost of repairs is much less, and no painting is required to preserve concrete structures subjected to ordinary usage.

Concrete structures may be erected with a rapidity and ease that are astonishing. Entire buildings have been erected in the time ordinarily taken to design and form into a whole the structural metal work for a similar building in steel.

Concrete has the power to resist the action of many chemical compounds that would cause the ultimate destruction of either wood or steel structures. The failure of wood construction is caused by the decay of the timber when exposed to the action of air combined with moisture or chemical acids. In order to preserve wood from this action, it must be covered with some resisting substance, such as sheet lead. This is expensive on account of the high first cost, and the cost of the lead burning required to make tight joints. This form of construction is not entirely satisfactory, as the lead, when exposed to the action of the gases, becomes brittle and soon cracks, allowing the chemical material to escape or become diluted and mixed with foreign substances.

Steel, when used for tanks or other structures containing chemical compounds, will corrode very rapidly, and must therefore be protected by lead or some other substance that will entirely prevent the steel and the chemical from

contact. In many cases this protection may be secured by a covering of concrete from one to three inches thick. If the steel be used to resist tensile stresses only, and the concrete to resist compression, the quantity of steel will be reduced, and the cost of construction with it. The resulting reinforced concrete structure will be equally strong and better able to resist corrosion. This method would require no extra steel to be added to prevent corrosion, as is now common practice in the design of steel structures in chemical plants.

Gases passing through a chimney from a steam boiler contain sulphur and other impurities which will act upon a steel stack in contact with them. Steel stacks as a consequence have a very short life. The use of reinforced concrete stacks is rapidly growing, because of their low first cost and their ability in resisting the action of the stack gases. An inner shell separated from the outer wall of the stack by an air space, should be provided to take care of the expansion caused by the heat of the stack gases.

It has been demonstrated by numerous experiments that concrete will protect steel from corrosion. From these experiments and from the examination of structures which have been taken apart, it may be concluded that concrete, when properly mixed and placed, gives the best protection yet discovered for the steel embedded in it, and that concrete may be safely used in all structures except where the chemical action would affect the concrete itself.

Durability. There is scarcely any class of manufacture which is not now being carried on in a reinforced concrete building. It is adaptable to any weight of loading, to high speed and heavy machinery, as well as to light machine tools, and to almost any style of design.

Recent scientific experiments, as well as actual experience, are favorable to the use of concrete under repeated and vibrating loads. The use of concrete in brackets for supporting crane-runs is an interesting example of severe application of loading. Several concrete buildings in San Francisco withstood the shock of the earthquake, while those around them of brick and stone and wood were destroyed.

While most materials tend to rust or decay with time, concrete under proper conditions continues to increase in strength for months or even for years.

Concrete expands and contracts with changes of temperature. Its coefficient of expansion—that is, its expansion in a unit-length for each degree of increase in temperature—is almost identical with that of steel, and on this account there is no tendency of the steel to separate from the concrete, and they act together under all conditions. As in building with other materials, provisions must be made, in long walls or other surfaces, for the expansion and contraction due to temperature, by placing occasional expansion joints or by adding extra steel. In factories of ordinary size, no special provision need be made, as the regular steel reinforcement will prevent cracking.

Fire-Resisting Properties. Reinforced concrete ranks with the best fire-proof materials, and it is this quality, perhaps more than any other, which is responsible for the enormous increase in its use for factories and other industrial structures.

Intense heat injures the surface of the concrete; but concrete is so good a non-conductor that, if sufficiently thick, it provides ample protection for the steel reinforcement, and the interior of the mass is unaffected, even in unusually severe fires.

For efficient fire protection in slabs, under ordinary conditions, the lower surface of the steel rods should be at least ¾ inch above the bottom of the slab. In beams, girders, and columns, a thickness of 1½ to 2½ inches of concrete outside of the steel, varying with the size and importance of the member, and the liability to severe treatment, is in general sufficient. In columns, whose size is governed by the loads to be sustained, an excess of sectional area should be provided.

One of the advantages of concrete construction as a fireproof material is that the design may be adapted to the local conditions. For example, in an isolated machine shop where scarcely any inflammable materials are stored, it is a waste of money to provide a thick mass of concrete simply to resist fire. On the other hand, for a factory or warehouse storing a product capable of producing not merely a hot fire —a hot, short fire will not damage seriously— but an intense heat of long duration, special provision may be made by using an excess area of concrete perhaps two or three inches thick.

The best fire-resistance materials for concrete are first-class Portland cement with quartz sand and broken trap rock. Limestone aggregate will not stand the heat so well as trap, while the particles of gravel are more easily loosened by extreme heat. Neither of these materials, however, if of good quality, need be rejected for building construction, unless the demands are especially exacting and the liability to fire great. Cinders make a good aggregate for fire resistance, but the concrete made with them is not strong enough for reinforced concrete construction, except in slabs of short span or in partition walls.

The fire resistance of concrete increases with age, as the water held in the pores is taken up chemically and is evaporated.

Table XX gives a comparative review of the insurance cost of industrial buildings made of concrete and those constructed of wood or brick.

Stiffness. A reinforced concrete building really resembles a structure carved out of a single block of solid rock. It is monolithic, or of one piece throughout. The beams and girders

TABLE XX

Concrete Factories vs. those of Wood or Brick, Approximate Yearly Cost of Insurance per $100

Exposures, none; area not large; good city department; no private fire apparatus except such as pails and standpipes.

Class or Structure	All Concrete		Brick Mill Construction or Open Joists		Wood Mill Construction or Open Joists		Add for brick or wood bldgs. in small towns and cities without best of water and Fire departments
	Bldg.	Contents	Bldg.	Contents	Bldg.	Contents	
General Storehouse	20c.	45c.	60c.	100c.	100c.	125c.	25c.
Wool Storehouse	20c.	35c.	40c.	60c.	75c.	100c.	25c.
Office Building	15c.	30c.	35c.	50c.	100c.	125c.	25c.
Cotton Factory	40c.	100c.	100c.	200c.	200c.	300c.	50c.
Tannery	20c.	40c.	75c.	100c.	100c.	100c.	25c.
Shoe Factory	25c.	80c.	75c.	100c.	150c.	200c.	50c.
Woolen Mill	30c.	80c.	75c.	100c.	150c.	200c.	50c.
Machine Shop	15c.	25c.	50c.	50c.	100c.	100c.	25c.
General Mercantile Building	35c.	75c.	50c.	100c.	100c.	150c.	25c.

NOTE—These costs are based on the absence of automatic sprinklers and other private fire protective appliances of the usual completely equipped building. They are not schedule rates, but may be an approximation to actual costs under favorable conditions based on examples in various parts of the country.

are continuous from side to side and from end to end of the building, while even the floor slab itself forms a part of the beams, and the columns are also either coincident with them or else tied to them by their vertical steel rods.

All this accounts for the extraordinary stiffness and solidity of a reinforced concrete strueture, and puts it into a different class from timber construction, where positive joints occur over every column; and even from steel construction, in which the deflection is greater.

COST OF REINFORCED CONCRETE

As a general proposition, reinforced concrete is almost invariably the lowest priced fireproof material suitable for factory construction. The cost is nearly always lower than that for brick and tile; and, with lumber at a high price, it is frequently even lower than brick and timber, with the added advantage of durability and fire protection.

In comparing the cost of different building materials, one must bear in mind that the concrete portion of the building is only a part of the total cost. Since the cost of the finish and trim may equal or exceed that of the bare structure, even if the concrete itself cost, say, 10 per cent more than brick and timber, the cost of the building complete may not be 5 per cent greater than with timber interior. The lower insurance rates will partly offset this, even if there is no other economical advantage for the fireproof structure.

The exact cost of a building in any case is governed by local conditions. In reinforced concrete, the design, the loading for which it must be adapted, the price of cement, the cost of obtaining suitable sand and broken stone or gravel, the price of lumber for forms, the wages of the laborers and carpenters, are all factors entering into the estimate. Reinforced concrete is largely laid by common labor, so that high rates for skilled laborers affect it less than many other building materials.

As a general proposition, it may be stated that the cost of reinforced concrete factories finished complete with heating, lighting, plumbing, and elevators, but without machinery, may run, under actual conditions, from 8 cents per cubic foot of total volume measured from footings to roof, to 16 cents per cubic foot. The former price may apply where the building is erected simply for factory purposes with uniform floor loading, symmetrical design—permitting the forms to be used over and over again—and with materials at moderate prices. The higher price will usually cover manufacturing buildings located in restricted districts, and where the appearance both of the exterior and interior must be pleasing. This does not include in either case interior plastering or partitions.

If we look at the side of the question which considers only the price of the raw materials in lessening the cost of a structure, we find that the cost of a cement structure can be lessened with respect to the cost of the cement which it contains, in two ways; (a) by employing a cheap quality; or (b) by using a small quantity.

A limit as to the quality has been mutually agreed upon by a large majority of cement manufacturers and users by the adoption of a standard specification for cement.

As to a minimum quantity, it may again be pointed out that the proper amount depends upon the characters of the other ingredients, upon the uses to which the completed structure is to be put, and upon the strength or quality required of the mixture under the governing conditions.

As to the relative costs of different mixtures of the various concrete ingredients, it is evident that the poorer the mixture (that is, the smaller the proportion of cement it contains), the less the cost per cubic yard. However, concrete of least cost may not be profitable under all circumstances.

If we consider the **cost of handling** the ingredients, the following points are worthy of notice: Make as much use of the force of gravity as possible. It is a common practice, on large works, to have high trestles and bins for the storage and handling of concreting materials. From these, the materials move downward by gravity through the mixers and into the conveying devices.

Another point is to have the mixing done as near the point of installation of the concrete as possible, since it is easier and cheaper to handle the dry material than the wet mixture. In the case of one large building with structural steel frame, a small electrically-driven portable mixer was used, and moved about in such a manner that it discharged directly at the point where the concrete was to be placed.

The **labor** item is another place wherein costs may be reduced—not necessarily by employing cheaper help, but by the economical handling of the men on the work. A few extra men who do nothing but the little odd jobs such as keeping planks in place for the men with the wheelbarrows can often save considerable time and money. The forming of a cycle of operations with each man having his particular part, is an ideal condition.

Where several thousand yards of concrete is to be raised to a considerable height, the use of platform elevators, bucket hoists, or derricks is recommended as a means of reducing the cost. The actual cost per cubic yard of hoisting concrete by each of these methods varies by only a few cents. With regard to other mechanical devices, it is sometimes found that the interest on their cost, together with their depreciation, more than offsets the saving in the cost of labor performed by them. In case of a bonus allowed for quick work, these devices are often used to advantage on account of their saving of time.

Every one at all conversant with the costs of reinforced concrete work knows only too well the disproportionate amount due to centering. In very heavy foundation work, this proportion may not be excessive; but in some buildings of reinforced concrete, the labor cost of installing and removing the wooden falsework, together with the cost of the material itself, has made up 50 per cent of the entire expenditure.

The cost of the centering material itself is heavy, and, when wood is considered, is growing heavier almost month by month. This is due to the constant and rapid increase in the prices of timber and lumber of all kinds, which has recently taken place.

A consideration of the detailed costs of concrete construction will show quite clearly where the temptations lie for slighting the work on the part of inexperienced contractors. As an example, take an eight-story office building which was recently erected in the East. In this building, which was 80 feet by 175 feet in ground plan, the percentage costs of the various items were as follows:

Labor	38	per cent
Cement	15	per cent
Stone and lime	9½	per cent
Steel	21	per cent
Lumber	10	per cent
Power	1½	per cent
Miscellaneous, unclassified	5	per cent

Table XXI shows a detailed consideration of the costs of reinforced concrete construction work. The steel cost given at the end of the table applies only to cost of putting in place and does not cover purchase price. These costs were obtained from a very large corporation engaged exclusively in reinforced concrete work and employing as superintendents and foremen experienced, skilled men. The average contractor handling occasional jobs cannot hope to reach these figures except under very favorable circumstances.

Insert Table

Table XXII gives dimensions and costs of a large number of representative concrete bridge structures.

TABLE XXI

Cost of Various Classes of Concrete Work

Kind of Structure	Average Cost of Forms per Square Foot				Average Cost of Concrete per Cubic Foot						
	Carpenter Labor	Lumber	Nails and Wire	Total	Concrete Labor	General Labor	Cement	Aggregate	Teams and Misc.	Plant	Total
	$	$	$	$	$	$	$	$	$	$	$
Building Walls (Above Grade) Average of 17 Structures	.085	.036	.002	.128	.090	.016	.073	.076	.025	.019	.301
Footing and Mass Foundations Average of 10 Structures	.057	.034	.002	.093	.045	.007	.071	.077	.007	.031	.239
Foundation Walls Average of 14 Structures	.068	.033	.002	.103	.076	.015	.080	.082	.019	.017	.289
Beam Floors of Reinforced Concrete Average of 18 Structures	.070	.045	.002	.116	.111	.020	.106	.083	.025	.034	.354
Flat Slab Floors Average of 3 Structures	.071	.038	.003	.111	.097	.009	.096	.070	.019	.024	.318
Concrete Columns Average of 9 Structures	.092	.036	.001	.130	.096	.027	.085	.049	.021	.023	.301
Concrete Slabs Between Steel Beams Average of 13 Structures	.061	.032	.002	.095	.102	.019	.128	.088	.024	.017	.359

Cost of handling steel for reinforcement (Average of 21 structures) was $8.52 per ton.

TIME FOR SETTING AND HARDENING

The time to be allowed for the necessary setting and hardening of the concrete before the forms are taken away, plays a very important part in the success of the structure. Mr. Edward Godfrey, in his volume on "Concrete," says:

"In counting the time that concrete should stand before removing the forms, those days when the temperature is at or below freezing should be counted out, or at least allowance should be made for almost total suspension of the hardening process during those days.

"It is safe to remove the forms from mass work receiving at the time no load except its own weight, in from one to three days: in warm weather with dry concrete, one day; in cold or wet weather or with wet concrete, more time. When the concrete will bear the pressure of

TABLE XXII

Dimensions and Costs of Concrete Arch Bridges.

PLACE	OVER	Total Length Bridge	Arch Spans	Width	Rise of Arch	Total Cost	Cost Per Sq. Ft.	Date Erected
Pine Road, Phila.	Pennypack Creek		2a 25'4"¾	34'3¼"	6'6"	$ 8,682	$ 4.94	1893
Neuhausel, Hungary	Neutra		6a 55'9"⅓	19'8"	3'8½"	13,700	2.00	1893
Richmond Av., Syracuse	Harbor Brook		12'0"	49'0"	2'6"	2,000	3.40	1894
Eden Park, Cincinnati	Park Ave.		70'0"	32'0"	10'0"	7,130	3.15	1905
Stockbridge	Housatonic		100 0"	7'6"	10'0"	1,475	2.00	1895
Belleville, Ill.	Richmond Creek		40'0"	53'0"	7'0"	10,500	3.90	1895
Topeka	Kansas River	693'0"	5a 125'0" 5a 110'0" 5a 97'6"	40'0"	......	126,000	4.50	1896
Green and Goat Islands	Niagara	271'0" 198'0"	4a 103'6" 1a 110'0" 4a 50'6" 1a 55'0"	40'0"	......	102,070	4.50	1900
Indianapolis	Fall Creek		6a 74'0"	65'0"	9'6"	106,340	3.70	1900
Washington	Rock Creek		80'0"	27'0"	15'0"	17,500	3.10	1901
Wabash Co., Ind.			25'0"	16'0"	......	573	1.45	1902
Forest Park, St. Louis			45'0"	45'0"	......	10,660	5.20	1902
I. C. R. R.	Big Muddy River	483'0"	3a 140'0"	34'2"	30'0"	126,000	7.60	1901-3
Plainwell, Mich.	Kalamazoo River	446'0"	7a 51'0"	26'0"	8'0"	19,900	1.80	1903
Saley St., Brooklyn	Prospect Ave.	125'0"	85'0"	60'0"	......	21,800	4.30	19—
Munich	Grunwald	720'0"	2a 230'0"	30'0"	42'0"	66,000	3.10	1904
Connecticut Ave., Washington		1,341'0"	2a 82'0" 5a 150'0"	53'0"	41'0" 75'0"	850,000	13.30	1905
16th St., Washington	Piney Branch	272'0"	125'0"	25'0"	25'0"	42,731	6.26	1905
Dayton	Great Miami	588'0"	660'-88'0"	56'0"	......	123,170	3.70	1904
Park Ave., Newark	Branch Brook Pk.	243'7"	132'0"	74'0"	10'2"	84,000	4.70	1905
Connington Viaduct		600'0"	10a 50'0"	16'0"	16'0"	52,206	5.48	1903
Greensburg, Ind.	Clifty Creek		80'0"	16'0"	12'6"	2,686	2.10	......
Wayne St., Peru	Wabash River	500'0"	6a 75' to 100"	30'0"	18'6"	36,900	2.46	......
Wabash, Ind.	Charley Creek		2a 75'0"	30'0"	18'0"	7,000	1.56	......
Muncie, Ind.			28'0"	16'0"	7'6"	640	1.21	......
Bangor, Me.		90'0"	46'7½" 23'2"	40'0"	8'7" 7'6"	6,500	1.80	......
Sandy Hill		1,088'0"	16a 60'0"	38'3"	......	73,000	2.00	......
Walnut Lane, Phila.		585'0"	233'0"	60'0"	......	[illegible]	7.85	......
Maryborough, Queensland		613'0"	a 80'0"	30'0"	4'0"	75,000	5.60	......
Lansing, Mich.	Grand River		130'0"	58'10"	36'0"	31,000	4.98	......

the thumb nail without indentation, it is ready to support itself in this class of work. Thin walls should stand two to five days. Slabs of reinforced concrete should stand about one to two weeks of good weather before being called upon to support their own weight. Slabs of long span may require more time than two weeks. At the same time the slab centering is removed, or even before it is taken down, the forms on the sides of beams and girders can be removed, leaving the supports of the bottoms in place for a longer time. This will afford an opportunity to inspect the surface of the beams and girders, and to plaster up any cavities before the concrete is too hard. Where practicable, it is well to leave the shores under beams and girders for three or four weeks. Large and heavy beams should be allowed to stand longer than short ones, because the dead weight is a greater fraction of the load they are designed to carry

"Column forms may be removed in a week or so, if the entire weight of the beams is supported by shores close to the columns; otherwise three weeks or more should be allowed.

"Arches of small span may have the centering removed in one to two weeks. Large arches should harden a month or more. Where practicable, it would be well to leave the concreting of the spandrel wall of an arch span until the arch ring has hardened and the forms are removed. The settling of the arch often cracks the spandrel wall, and gives an unsightly appearance to the bridge.

"Ornamental work should have the forms removed as soon as possible, so that defects can be plastered up, and so that swelling of the wood will have less time to act.

"Falsework should be removed carefully, without jar to the concrete either by hammering on the boards or by dropping heavy pieces on the floor below. The supports should not be removed when any unusual load is on the floor. Materials should not be stored on floors that are not thoroughly hardened and self-supporting.

"Concrete reinforced work should ring when struck with the hammer, before the supports are removed."

GENERAL PRINCIPLES OF REINFORCED CONCRETE DESIGN

The use of some form of steel reinforcement has already been shown to be a necessity in any structure where tensile stresses will be developed. It remains for us now to find out where to locate the steel reinforcement, and how much is needed for a given load upon a member.

In the case of a concrete beam or girder, the horizontal steel reinforcing rods should be located as near the bottom of the beam as possible, still allowing sufficient thickness of concrete underneath to protect them in case of fire or exposure to liquids, gases, or other agents tending to cause corrosion. This is, as has already been stated, on account of the stresses in the bent beam being divided on each side of an imaginary line located near the center of the beam section (for rectangular beams) and called the **neutral axis.** The stresses above this line are compressive stresses, and are taken care of by the strength of the concrete itself; while those below the line are tensile, and, on account of the weakness of concrete in tension, are take care of by the steel.

This reinforcing steel may be in the form of

rods or **bars. unit-frames,** or **structural shapes,** for girders, beams, and long spans of construetion work, or for short spans bearing heavy loads; but the floors, roofs, etc., of short spans and light loads, the **sheet fabrics** can often be used to advantage.

At the part in the span where the greatest bending action occurs, the depth of the reinforcement below the top surface of a beam or girder varies with designers from 7/8 to 10/11 of the depth of the beam section. This allows a good protection against fire in ordinary sizes of beam section, and also allows for plenty of concrete around the metal to resist shearing action along the rods or to prevent the slipping or pulling-out of the rods.

Godfrey, in his work on "Concrete," recommends that the spacing of rods or bars in beams, when the diameter is 1/200 of the span, should be four diameters for square rods, and three diameters for round rods. If the rods are of smaller diameter than 1/200 of the span, the spacing may be closer. The distance from center of outside rod to side of beam should be one-half the spacing.

The web of a beam or girder is often protected at the end against diagonal tension cracks, by bending up the end thirds of the rods and bringing their ends nearly to the top surface of the beam. These ends are then fitted with anchors which hold them firmly in the concrete. This construction has been found to strengthen the web of the beam greatly, and to cause it even to approach the condition of an arch in resisting loads. Professor Talbot found, as a result of many tests, that the loads carried by beams with all the reinforcing bars bent up but not anchored did not differ much from those in cases where the bars were all straight. The failure was slower in the case of the bent bars, and warning of approaching failure was given. He found, however, that if part of the bars were left straight and alternated with the bent ones, the web was considerably strengthened.

These tests also showed that the use of U-shaped stirrups generally strengthened the webs of beams, but the amount of additional strength depended largely upon the quality of the concrete. The stirrups did not exert any considerable strength, however, until a diagonal crack had formed in the beam.

Mr. Ransome's rule for placing the stirrups in a beam is to place the first one at a distance from the end equal to ¼ the depth of the beam; the second, a distance of ½ its depth beyond the first; the third, a distance of ¾ the depth beyond the second; and the fourth, a distance equal to the depth of the beam beyond the third.

The main point in putting in this steel reinforcement is to get just enough steel below the neutral axis to balance the strength of the concrete above it. We do not wish to use too large sizes of rods in light beams, as the result would be a crushing of the concrete on the top side of the beam, or a shearing along the rods, when the beam was bent. Professor Talbot has shown from his experiments that steel rods whose combined area amounted to from one to one and one-half per cent of the area of the beam section above a line drawn through the center of the rods (the half-holes above this line being figured as a part of the beam section) would allow the beam to fail by tension in the rods. For percentages of steel higher than one and one-half, there is a liability to failure by crushing of the concrete.

These experiments were performed upon beams composed of 1:3:6 concrete. A 1:2:4 concrete would permit of a slightly larger percentage of steel, on account of the greater compressive strength of the concrete.

Tests for adhesion have also shown that if rods of a greater diameter than 1/200 of the length of the beam are used, they are liable to pull out of the concrete without breaking. This is due to the want of proper adhesion between the rod and the concrete, in comparison with the strength of the rod itself. Tests have also shown that beams whose depth is greater than 1/10 the length, or span, when reinforced with horizontal rods, or with rods bent over at the ends and not provided with anchor plates, will fail when loaded to the breaking point, in a manner similar to that shown in Fig. 47. This is due to diagonal tension produced in the beam as a result of the combined vertical and horizontal shear together with the direct tensile stress.

The theories regarding flexure in reinforced concrete beams are based upon mathematical principles, as well as the principles of mechanics. Therefore the detailed design of any important reinforced work—especially when comparative costs must be considered and where failure would result in serious disaster to property or life—should be under the immediate supervision of an engineer trained in these principles and competent to apply them wth judgment to the work in hand.

SIMPLE, PRACTICAL RULES FOR DESIGN

Several empirical formulas, or working rules based on experience and observation in actual practice as distinguished from refined theoretical calculations, have been suggested for different types of design, and their authors claim that they closely follow the results of reliable tests. These formulas should be used with care, and applied only to the class of work that supplied the data from which they were derived.

Mr. Homer A. Reid, in his volume on "Concrete and Reinforced Concrete Construction," presents two simple, approximate working formulas for the design of beams. These are **Wason's formula and Ransome's formula.**

Wason's Formula. This formula is "based on the following assumption: that there is a perfect bond between the steel and the concrete within the limits of the working stresses of the combination. That the steel takes the entire tensile stress and the concrete the entire compressive stress. That the neutral axis is assumed to be half-way between the center of the reinforcing bars and the top of the beam. That the center of pressure of the concrete under compression is considered as being two-thirds of the height from the neutral axis to the top of the beam. The distance from the center of pressure of the concrete in compression to the center of the reinforcement, equals $5/6$ of the distance from the top surface of beam to the reinforcement."

d = Effective depth of beam (top to reinforcement).

l = Span in inches.

F_s = Total stress in steel.

W = Total uniform load in pounds.

Then, taking the center of pressure as the center of moments, the **resisting moment**

$$M = \tfrac{5}{6} d F_s.$$

The **bending moment** of a freely supported beam under a uniformly distributed load is

$$M = \tfrac{1}{8} Wl.$$

Equating these two moments, and solving for F_s, we obtain:

$$F_s = \frac{Wl}{6\tfrac{2}{3} d}.$$

Example. Determine amount of steel required for a beam of 12½ ft. span to carry a total uniform load of

12,500 lbs., assuming an effective depth of 14$\frac{4}{10}$ inches, and using a unit-stress for the steel of 16,000 lbs. per square inch.

Since 12½ ft. = 150 inches, we have:

$$F_s = \frac{12{,}500 \times 150}{6\frac{2}{3} \times 14\frac{4}{10}} = 19{,}500 \text{ lbs.}$$

$$\text{Area of Steel} = \frac{19{,}500}{16{,}000} = 1\tfrac{22}{100} \text{ sq. in.}$$

Two bars $1\frac{3}{16} \times 1\frac{3}{16}$-in. give an area of $1\frac{32}{100}$ sq. in.

After determining the total stress in the metal, the area of the reinforcement is determined by dividing the total stress by a safe working stress to determine the area of metal. Bars of proper size are selected to make up this area, a convenient spacing selected, and the area of the concrete adjusted to resist the compression. Mr. Wason uses 16,000 lbs. per square inch tension on the steel; and for a 1:3:6 concrete, an average of 500 lbs. per square inch in compression on the concrete; and requires 32 square inches of concrete in the upper third of the beam for each square inch of steel (in the lower part). This averages very nearly 1 per cent of reinforcement.

The above ratios are applied to the use of Ransome twisted bars, which have a high elastic limit and give a factor of safety of about 4.

In the above problem, the total compression is 19,500 lbs.; this, divided by 500 lbs., gives a required area in the upper third of the beam of

39 square inches. 39×3 = 117 square inches; total area of beam, 117 square inches, divided by $14\frac{4}{10}$ inches depth assumed, gives $8\frac{13}{100}$ inches width of beam. A width of $8\frac{1}{4}$ inches may be used.

Ransome's Formula. **Ransome's** formula for a simple beam uniformly loaded is:

$$S = \frac{Wl}{7d}$$

in which,

W = Total dead and live load, in tons.
l = Span, in inches.
d = Depth of steel below top of beam = Effective depth.
S = Maximum stress in beam, either tension or compression.

When the beam is not uniformly loaded, the formula becomes:

$$S = \frac{BM \times 8}{7d},$$

in which BM equals the maximum bending moment in inch-tons.

In order that the compressive stress per linear foot of width resulting from a chosen value of d shall not exceed the safe compressive strength of the concrete, there must be 16 square inches of concrete above the bars for each ton of stress.

$$16S = 12d,$$

from which,

$$S = \frac{3}{4}d.$$

Substituting this value of S in the above formula, we have:

$$\frac{3}{4}d = \frac{Wl}{7d}$$

$$d = \sqrt{4Wl}$$

Having obtained **d**, the total stress in tons, S=¾ d.

Example. Assume a flat floor slab, having a span of 12 feet carrying a live load of 150 lbs. per square foot.

It is necessary to assume the dead weight of the floor. Let this be taken as 75 lbs. per sq. ft., making a total load of 225 lbs. per sq. ft. The total load W in tons on a strip of floor 1 ft. wide would be:

$$\frac{12 \times 225}{2{,}000} = 1\tfrac{35}{100} \text{ tons,}$$

and we have for **d**,

$$d = \sqrt{\frac{4 \times 1\tfrac{35}{100} \times 12 \times 12}{21}} = 6\tfrac{8}{100} \text{ inches.}$$

The total stress in the bars would equal $\frac{3}{4} \times 6\frac{8}{100} = 4\frac{56}{100}$ tons. Assuming an allowable working stress on the metal of 8 tons per sq. in., there will be $\frac{57}{100}$ sq. in. of metal required—or four rods ⅜ in. square—in each foot width of slab. When ¼-in. rods are used, the distance from center of rod to bottom should be at least ½ in.; and ¾ in. for ½ in. square rods. For

⅜-in. rod reinforcement, we will have a total thickness of 6¾ inches.

In calculating the beam dimensions and amount of reinforcement for ribbed slabs, the formula

$$S = \frac{Wl}{7d}$$

is used. This condition, however, is imposed, that the upper third of the beam, including the flat slab connecting the ribs, shall contain at least 5 sq. in. of concrete for each ton of stress given by the formula. This condition prevents the concrete in the top of the slab from being strained beyond its safe compressive strength.

In the design of an ordinary beam which is to rest freely upon a support at each end, we have only to consider tension as occurring in the material on the under side of the beam. This is not the condition generally found in reinforced structures. The beams, girders, and floors, commonly form a continuous mass, thereby fixing the beams and girders more or less firmly at the ends. This condition prevents the ends of the beam and girders from inclining, as the bending occurs in the center, thereby causing the top surface of the beam to come into tension over the supporting walls or columns. Reinforcing rods must be placed near the top surface of such beams and girders, and anchored firmly, just as they are placed near the bottom surface in the middle of the span.

In floor construction where some form of wire fabric is used as a reinforcing agent, the sag of the fabric as it is stretched continuously across several spans brings it toward the lower surface of the floor slab in the center, thereby giving it reinforcement in the proper place; and the fabric slants upward again toward the ends of the span in order to pass over the beams into the next span. This upward slope of the material again places strength where it is needed.

When the ends of rods are to be joined in a continuous construction, unless they are joined by some form of connection, practice shows that they should overlap in the concrete for about fifty diameters.

In the case of **T-beams** of reinforced concrete, the assumption upon which calculations are often based is that we may substitute for the actual T-section the area of the rectangle found by extending the sides of the flange section downward until they meet a horizontal line passing through the center of gravity of the end sections of the reinforcing rods in the lower part of the web, and we may then figure as in the case of beams of rectangular section. There is claimed to be but small error in this assumption.

The principles which govern the **design of reinforced columns** vary somewhat from those used in the design of beams and floors. Shear plays a prominent part in the failure of a concrete column. When a plain concrete column fails, the general tendency is one of bulging, and sliding of the concrete in the bulged part. Longitudinal rods of small diameter, when placed near the surface and not connected, do very little to prevent this action, on account of the bending of the rods. Longitudinal rods tied or bound together by bent bars or heavy sheet fabric so as to form a stiff cage surrounding the core of the column, will prevent this buckling action. Many engineers prefer a round or square bar construction to the flat hoop form, on account of better bond in the concrete, especially if the space between the flat bars is small.

We have already referred to the possible tendency in the column to bend when subjected to eccentric loading. If the center of gravity of the applied load at the top of the column does not exactly coincide with the center line of the column, the compression increases on one side of the column and grows less on the other. The limiting boundary within which the center of gravity of the applied load can act, and yet not put one side of the column into tension, is the middle third for a square-section column, and the middle fourth for a round-section column.

Besides the weakening effect of the tension in such a case of loading, the shear (another weakness in concrete) is increased on the compression side of the column. When longitudinal bars bound together as just described are used, the flexural stresses and the tendency to bulge are resisted.

Godfrey suggests that a round or octagonal

column with a coil of square steel rods $\frac{1}{40}$ the diameter of the column, the coil itself $\frac{7}{8}$ the diameter of the column, bound together by wiring to the inside of the coil eight longitudinal rods of same size material as the coil, will produce good results. For lengths of columns up to ten diameters, 550 lbs. to the square inch is recommended as the allowable unit compression in the concrete. For lengths between ten and twenty-five diameters, he suggests that the formula

$$p = 670 - 12\frac{l}{D},$$

in which l is the length of the column in inches and D is its diameter in inches, will give the allowable pressure per square inch in the section. Twenty-five diameters should be the limiting length of column.

In columns which carry heavy loads, as in high buildings, a more suitable construction is that of a steel framework, built up of channels or angles and lattice-work, which is then filled with and surrounded by concrete.

General Design of Reinforced Concrete Buildings.—In all concrete work, the essence of economy is found to lie largely in the proper proportioning of ingredients according to the materials available and the work in hand, and in simplicity and duplication of forms. The designer—especially of a reinforced structure—should therefore eliminate as far as possible all projecting members, such as beam panels, cornices, belt-courses, and offsets.

Owing to the comparative youth of this type of structure, it is only within very recent years that architects and engineers have given any very great attention to the development of designs essentially suitable to reinforced concrete. It is the general practice to design in brick, stone, and steel, and then to call upon a reinforced concrete engineer to reproduce a structure in reinforced concrete. This is an imperfect and unscientific method, and unfair to the development of the true value of the concrete structure. Many architects and engineers are, as yet, too unfamiliar with the characteristics of concrete to design a structure solely from that point of view. They think in terms of brick and steel, stone and wood, and design in these materials, and then attempt to adapt concrete to their structure. The architect who would design intelligently for reinforced concrete must think in terms of reinforced concrete, which possesses peculiar characteristics essentially its own. Any design which is to be carried out in this material should be adapted to its characteristics and qualifications. The reinforced concrete building must be essentially a plain building, and its architectural effect must be developed by the study of the relation of the openings to the masses, together with the assistance of some surface finish suitable to the problem, or some color scheme, applied by the insertion, in the face of the structure, of tile, pebble, or mosaic work.

REINFORCING MATERIALS AND SYSTEMS.

Steel is the common medium used in reinforcing concrete. Although the ways and forms in which it is used are varied, and allowing that each form may have its special advantage, the general principle of placing the metal where it will take up the tensile stresses is common to all forms.

Some of the more common methods of application are as follows: by steel bars, rolled shapes, sheet metal, expanded metal, woven wire, plain and barbed wire, old structural work of various kinds, wire rope, chain, etc.

Steel bars are divided into different classes—plain bars, deformed bars, and trussed bars; plain bars comprise the ordinary round, square, and flat bars; while deformed bars are illustrated in Figs. 51, 52, and 53, and in Plate 6.

Examples of trussed bars are shown in Figs. 54 and 55.

The round bar is generally preferred to either a square or flat bar, on account of the liability of cracks in the concrete developing in setting when the sharp-cornered bars are used. The round bar also has the advantage of being easily threaded and fitted with nuts at the ends, in case plates for prevention of slipping of rod in

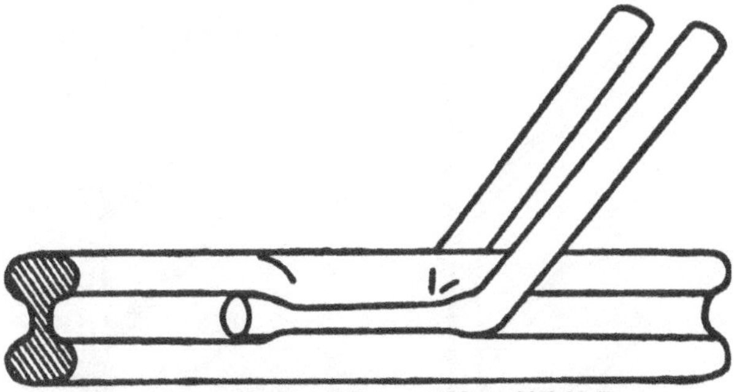

Fig. 51. Deformed Bar—"Monolith" Type.

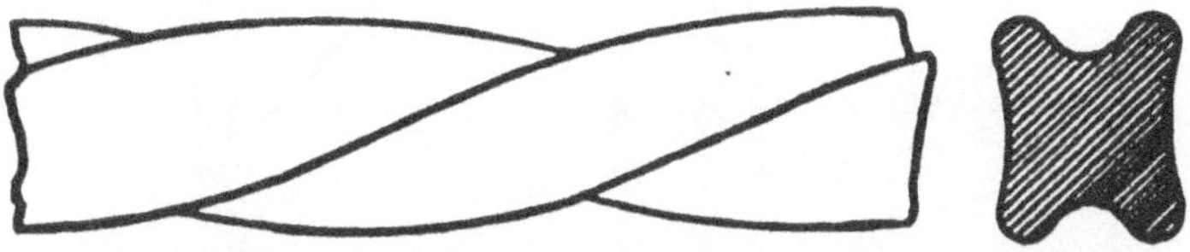

Fig. 52. Deformed Bar—U-Twisted Lock Type.

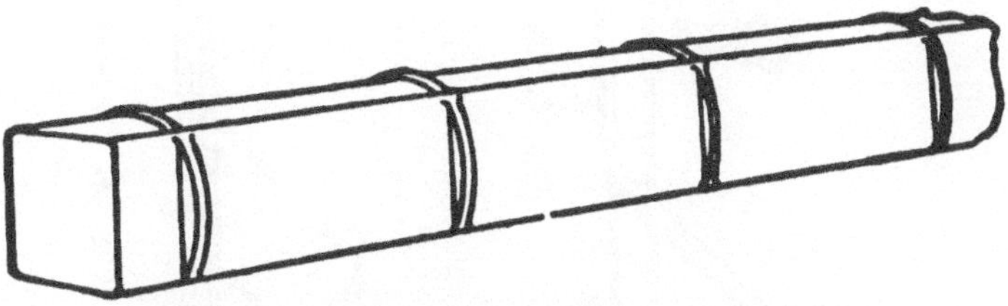

Fig. 53. Deformed Bar—Square Lug Type.

the concrete are to be used at the ends of the bar.

In trussed bars, the diagonal arms prevent the lower main part of the bar from slipping, and strengthen the beam or girder against the effects of internal stresses throughout its depth. In ordinary forms of deformed bars, the primary object in each is to provide a better bond against slipping between the bar and the concrete. The

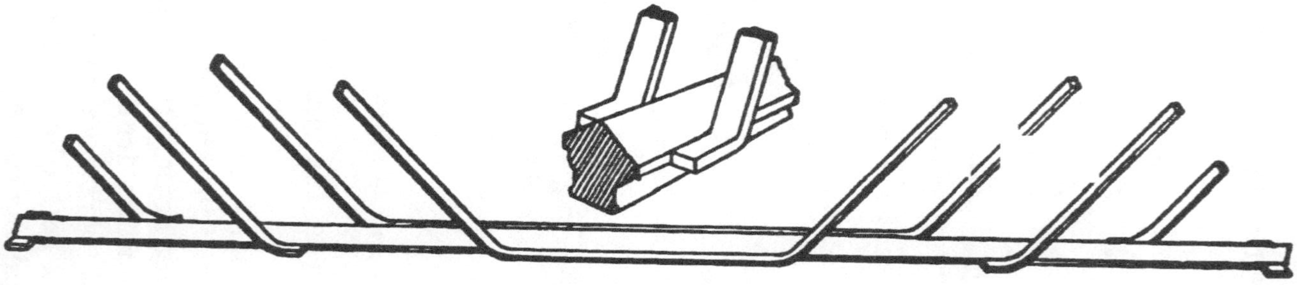

Fig. 54. Trussed Bar—Kahn Type.

Fig. 55. Trussed Bar—Herringbone Type.

relative merits of the various bars will not be discussed; but according to tests conducted by Professor Talbot, it is questionable whether there is really any great gain in strength effected by the use of the ordinary deformed bar.

Column hoops are formed by bending either flat, square, or round bars into a spiral which is held in shape by longitudinal stays having various means of holding the metal to the stay. An eccentric load placed upon a column tends to throw one side into compression, and the other into tension. Therefore columns need near-surface reinforcement, just as beams when subjected to similar stresses.

Steel shapes, possibly more commonly of the I, T, channel, and angle forms, are used extensively in floors, girders, columns, and even in heavier foundation work.

Sheet metal, expanded metal, metal lath, woven wire, etc., are used in roofs, ceilings, sidewalks, concrete piles, partitions, short-span floors, columns, ornamental work, or, in fact, in almost any construction where there are not too great stresses, or where flexibility within the forms is needed. Figs. 56 to 62 show some of the various forms of these materials.

It is not uncommon in light work to find ordinary **strong wire**, and often **barbed wire**, a very valuable reinforcing agent. In rough foundation work, **scrap iron** is often used as a bonding material; but the use of cast iron, whose tensile strength is exceedingly low, would hardly be recommended in any place where bending might occur. Broken **machine parts** and old iron **hoops** from barrels, are often used on the farm in rough construction work. Old **steel rails** are very conveniently used in footings, but they should be clean and free from oil

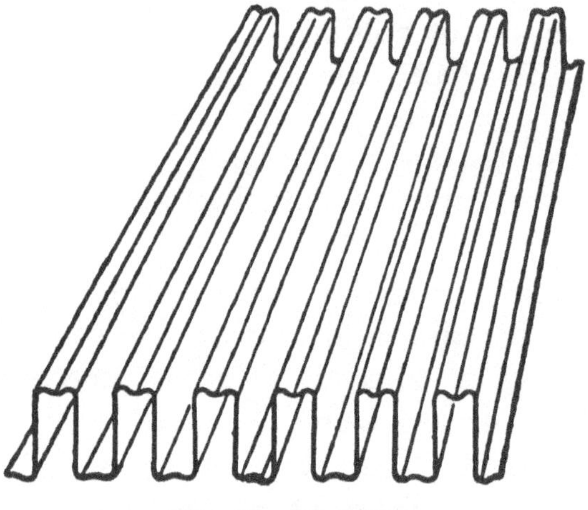

Fig. 56. Steel Plate

or heavy rust. In fact this requirement should apply to all material which is to be embedded in concrete. A little rust is sometimes beneficial in helping to remove the mill scale from the new bars thus allowing the concrete to obtain a better hold.

Wire rope or **cable** is an unsatisfactory form of reinforcement, not only on account of its flexibility, but also on account of the stretch resulting when under stress. Tests have shown a

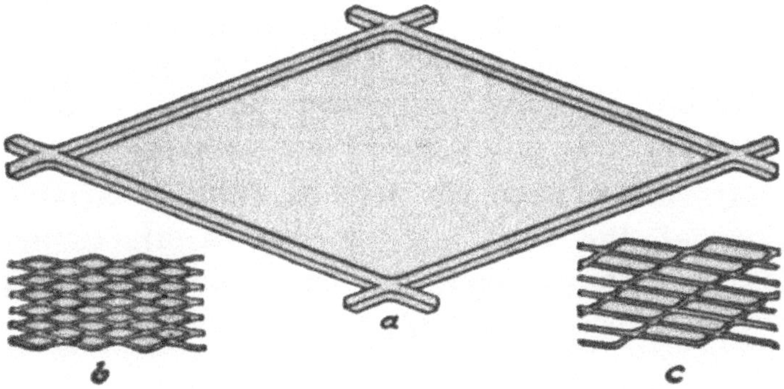

Fig. 57. Expanded Metal of Standard Mesh.

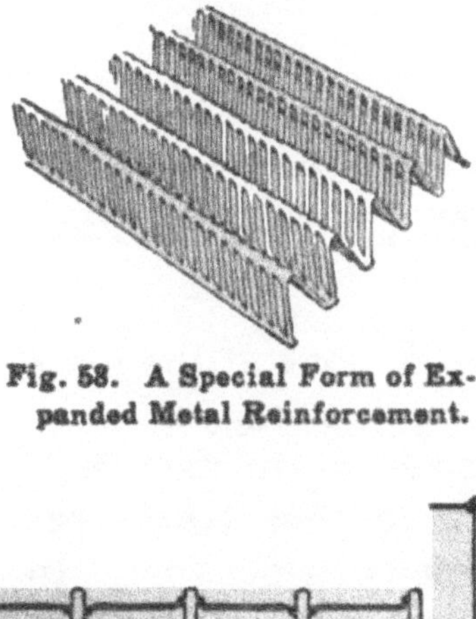

Fig. 58. A Special Form of Expanded Metal Reinforcement.

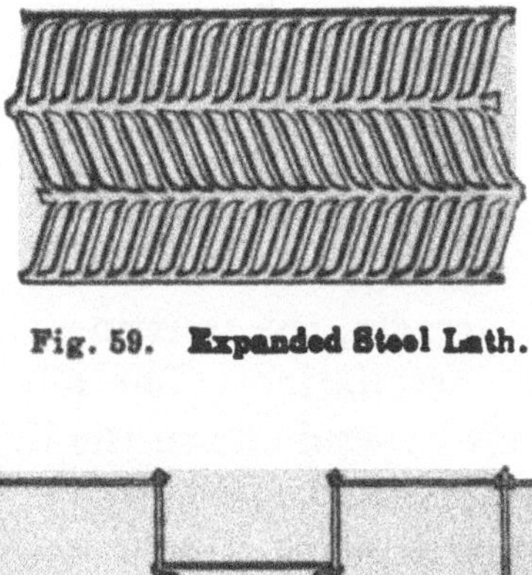

Fig. 59. Expanded Steel Lath.

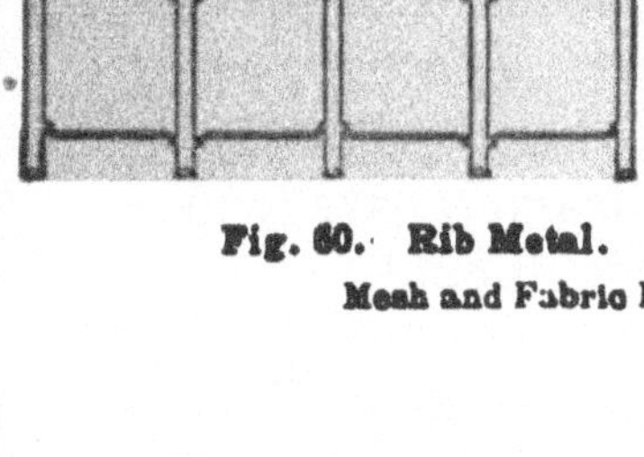

Fig. 60. Rib Metal.

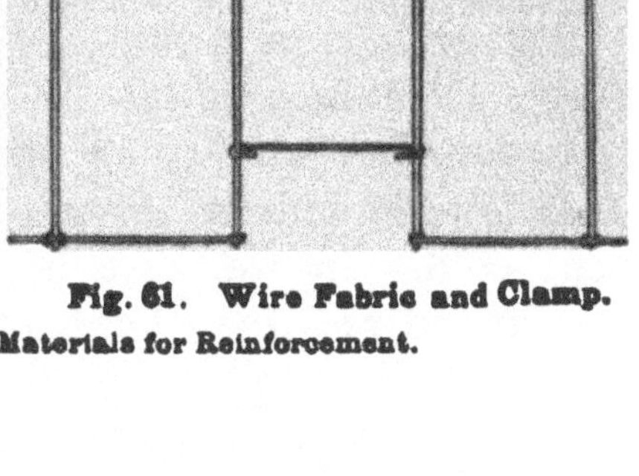

Fig. 61. Wire Fabric and Clamp.

Mesh and Fabric Materials for Reinforcement.

wire cable to stretch four times as much as a plain steel rod under the same unit-stress. Therefore a beam reinforced in this manner would probably crack under low stresses.

Bars for reinforcing may be bent or twisted either at the work or at the manufacturing plant. Twisted bars can be readily obtained from the manufacturers; but in cases of large

A B

Fig. 62. Lock-Woven Fabric of Standard Gauge

A—Long Wires, No. 10 gauge, on 4-in. centers; cross wires, No. 9 gauge, on 6-in. centers. B—Long wires, No. 3 gauge, on 4-in. centers; cross wires, No. 10 gauge, on 6-in. centers.

contracts, the twisting is often done at the place of installation. The number of twists per linear foot depends upon the diameter of the bar; thus, for ¼-inch bars, there may be five twists per foot; and for 1-inch bars, one twist per foot. In computing the cross-sectional area of steel in reinforced concrete, the twisted bars are figured as square bars of the dimension before twisting.

Fig. 63 shows a pair of tongs which may be used in bending light steel bars during the construction work. Fig. 64 shows a power bender used for the heavier shapes. Light bars may be bent cold. In the case of heavier bars which are heated and then bent, care should be taken to see that the steel is properly annealed afterwards, to restore its original properties.

Life of Steel in Concrete. Good concrete is one of the best known preservatives of steel. The mixture should be wet when the steel is embedded, and thoroughly tamped so as to coat the steel completely. Dry or porous mixtures

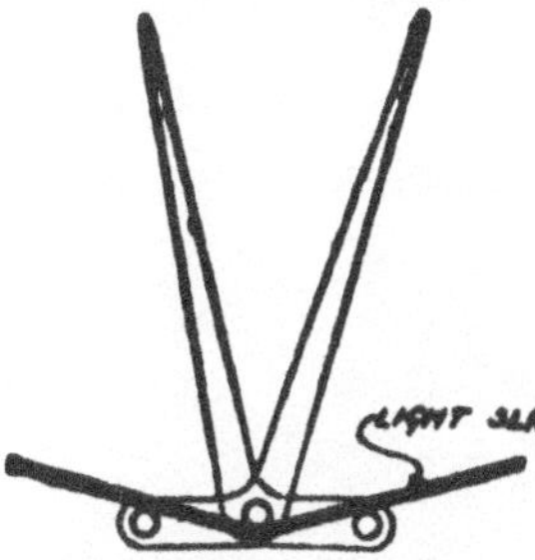

Fig. 63. Tongs for Bending Light Steel Bars.

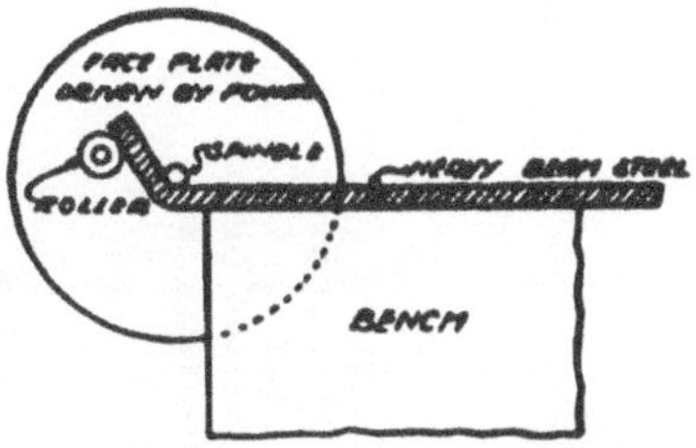

Fig. 64. Power Bender.

allow access to the steel of air or other vapors, which should be avoided. It is generally assumed that steel embedded in concrete does not corrode. Instances where steel has stood in concrete for years seem to prove this. In fact, it has been stated that steel with a slight coating of rust, when placed in position, has been found to be bright when removed after considerable lapse of time. An exception might be made to the above statements in case of injury to the concrete causing cracks, since rust or corrosion is liable to occur at these cracks.

An objection has been made to the possible corrosion of steel embedded in cinder concrete through the action of sulphur in the cinders; but experiment seems to show that such is not the case, and that the corrosion resulted mainly from the rust, or iron oxide, in the cinders. Cinder concrete should be especially well rammed while wet.

SYSTEMS OF REINFORCEMENT

Although we have referred in a general way to the use of various forms of steel as reinforc-

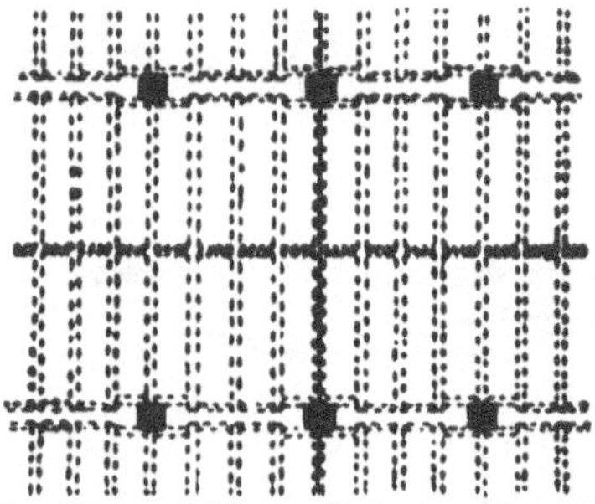

Fig. 65. Plan of Floor Reinforced with Rods

ing agents, architects and engineers have begun to use these materials in more or less advantageous grouping of parts. The following systems illustrate a few of the combinations which are in use in building construction to-day. Most of these are the results of careful computations by competent men, and some have been thoroughly tested by severe conditions in construction work.

Reinforcement for Floors, Beams and Girders. Fig. 65 shows one of the pioneers in the line of reinforcing concrete floors and beams by the use of rods. This system consists of reinforced beams, quite thin and deep, with web stiffeners, spaced about like floor joists in ordinary construction work, and supported by girders at the ends. Several bars are grouped in parallel lay-

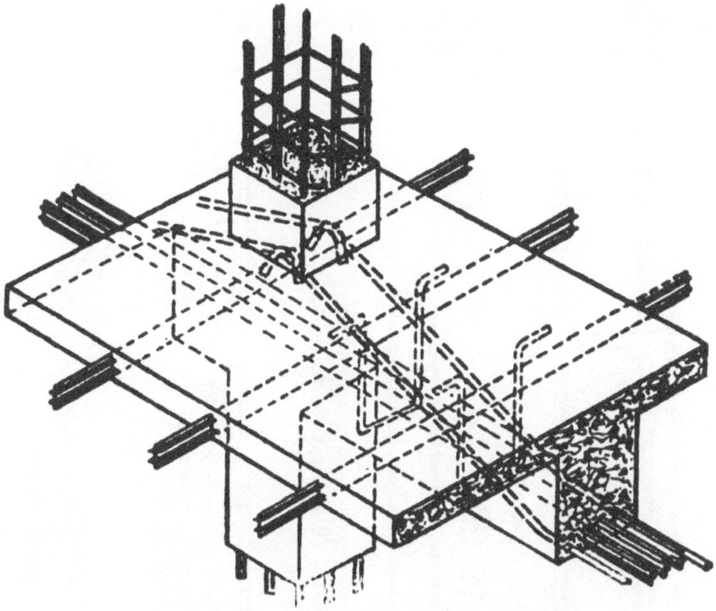

Fig. 66. Reinforced with Special Bars

ers for the heavier girder reinforcements. In the case of long spans in the beams, a central stiffening web is often used as bridging between the beams.

A system employing a form of trussed bar, as shown in Fig. 54, is often used. This special form of trussed bar is made of a grade of medium open-hearth steel with an elastic limit up to

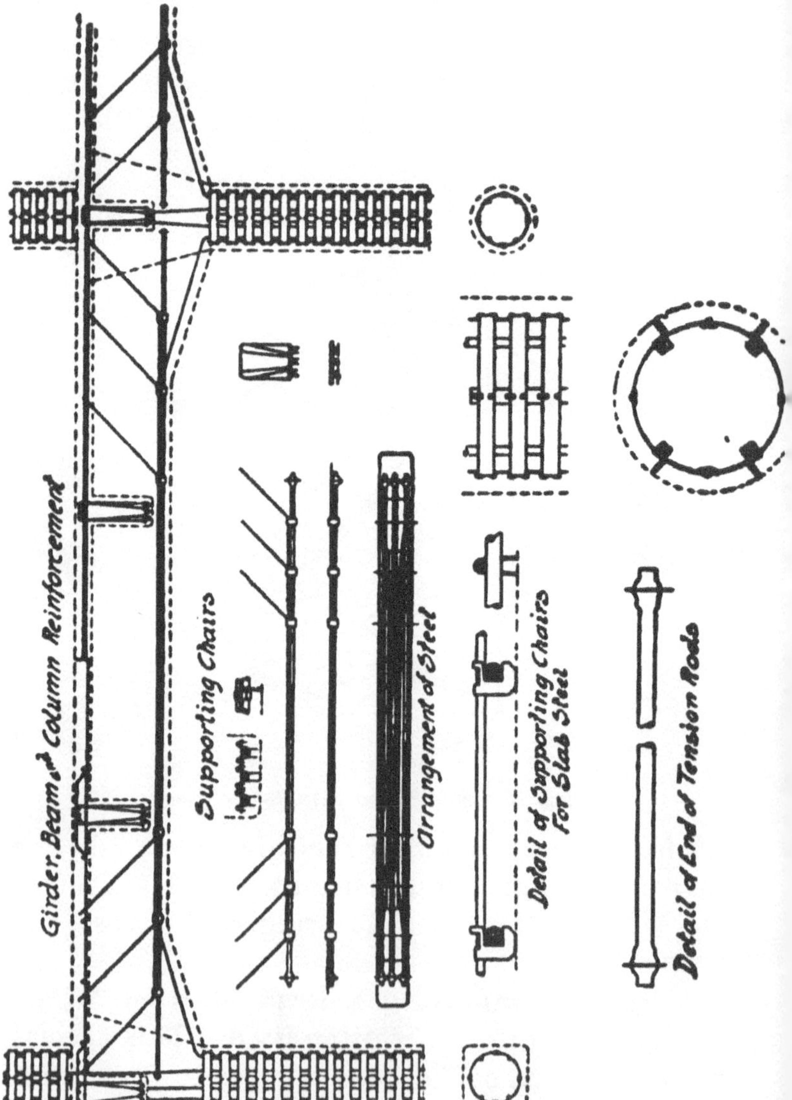
Girder, Beam and Column Reinforcement
Supporting Chairs
Arrangement of Steel
Detail of Supporting Chairs For Slab Steel
Detail of End of Tension Rods

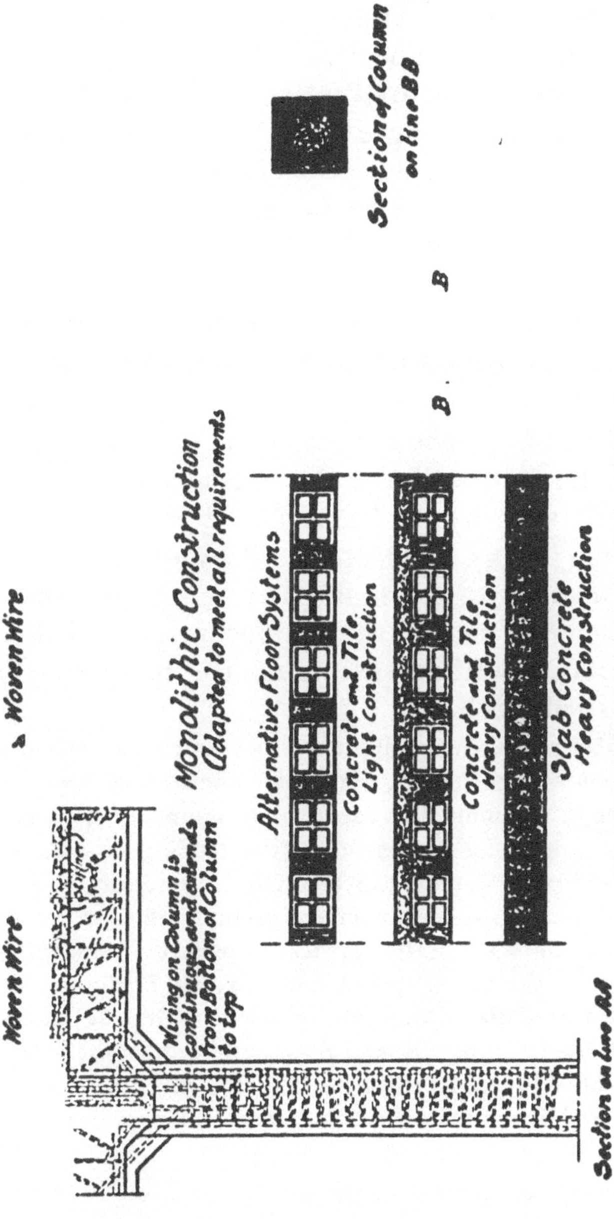

Fig. 88. Girder, Column and Floor Reinforcement.

42,000 pounds and an ultimate tensile strength of 70,000 pounds. The cross-section illustrated has two horizontal flanges or wings projecting at diametrically opposite corners. These winged portions are sheared up at intervals, and bent so as to make an angle of 45 degrees with the main portion of the bar. It is claimed that in this bar there is no waste metal at any point, and that proper reinforcement is provided at every place it is needed.

Figs. 66, 67, and 68, show three other systems of special construction, each one of which has its own special merits.

The "Mushroom System" is so-called from the peculiar formation of the rods around the column head, and from the remarkable rapidity with which it may be erected. The idea of this system is primarily to simplify the centering, and thus reduce the cost of the temporary part of the construction without skimping the materials in the finished work.

The arrangement of the reinforcement is designed with a view to securing the maximum efficiency of the materials through straining the concrete in a number of directions, the compression in one direction tending to balance and offset that in another; incidentally to concentrate the maximum amount of reinforcement around and over the support where the shear is the greatest; to enable removal of the forms at the earliest possible period; and finally to eliminate beams and ribs which interfere with light and catch dust, cost money to plaster and finish, and reduce the clear story height. The flat ceiling so obtained gives free and unobstructed illumination from the windows: it permits one to place partitions anywhere without regard to the floor, giving an unusual stiffness and solidity due to the fact that a part of the material which in the beam type is placed in the ribs, is consolidated in the slab, making the slab of unusual thickness with an actual decrease in the total amount of material where the loads are at all heavy.

The **"Spider-Web" System** of reinforcing is shown in Fig. 69.

In the Spider-Web floor and ceiling construction, all rods are small—usually three-eighths inch to one-half inch in diameter, concealed in a continuous slab, and therefore fire-

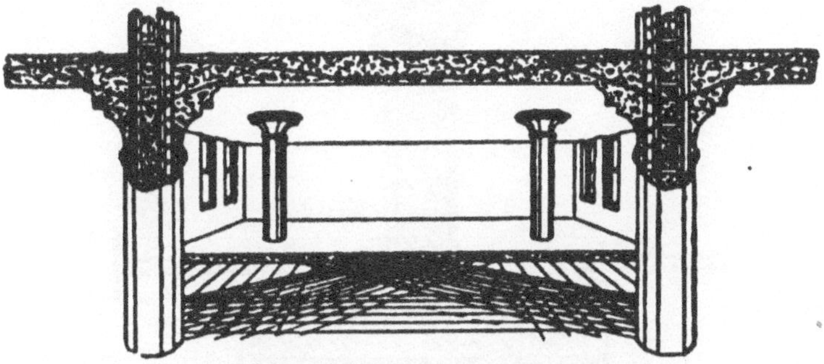

Fig. 79. Spider Web System.

proof. The octagonal hoop, which acts as a key to the eight trusses, is eight feet across. In buildings where columns are spaced twenty feet to centers the distance between column heads is twelve feet. The steel trusses comprising column head are usually formed of three-quarter inch round rods, and all pass through holes in the vertical steel forming core of the column. As compared with ordinary beam construction, not only is the span, in effect, reduced, but the carrying capacity is further increased, because the spider-web pulls from all angles when loaded, whereas the beam construction pulls from only two directions.

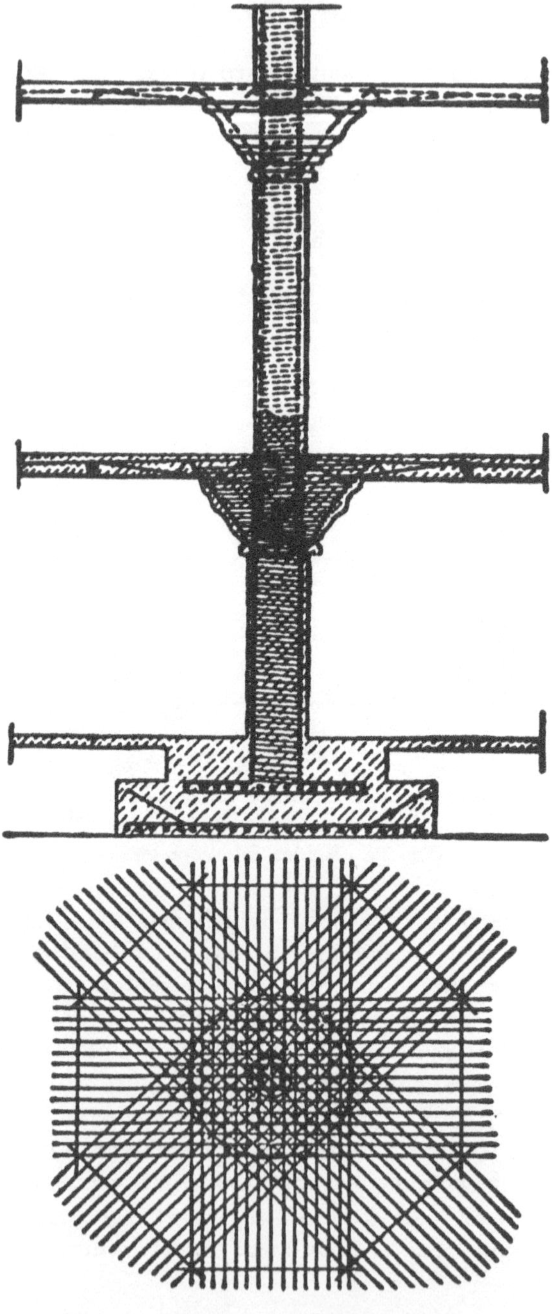

Fig. 70. Details of Umbrella System

The "**Umbrella" System** is shown in Fig. 70.

This system is claimed to be a departure from other systems of this genreal class in that the columns are continuous, and, if spliced, a **telescope splice** is used in the umbrella head, which has triple hooping reinforcement. Not only is the value of the umbrella head thus increased threefold, but the column loads from above are substantially transmitted into the center of the column below, thereby insuring a continuous column and minimizing the danger of eccentric loading of the lower column.

As an example of the **Unit-Beam System** of reinforcement, the **Pin-Connected Girder** Frame, as shown in Fig. 71, is given. These frames are shipped from the manufacturers ready to be lifted into the forms. No expensive field work, either blacksmithing or assembling, is required. Each frame is so numbered that the steel is erected rapidly and economically; and the dangers of misplaced steel, which always attend the use of loose rod reinforcement, are entirely avoided. The mechanical features pointed out as of greatest importance are:

Diagonals rigidly attached at both ends, which may be spaced as frequently as is necessary to resist the shearing stresses:

Carrying one of the main members to the top at the supports, and returning it to provide for negative moments;

A link and pin connection over each point of support, giving each frame a mechanical connection with adjoining frames, so that bonding action of the concrete is not depended upon to transmit stresses from beam to beam;

In use as beam reinforcement, the required amount of steel is made up by using as many frames (units) as are necessary.

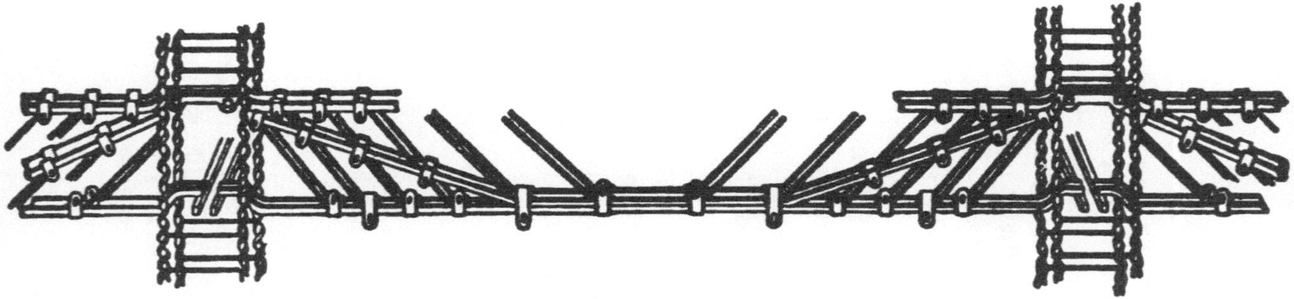

Fig. 71. Pin-Connected Girder Frame.

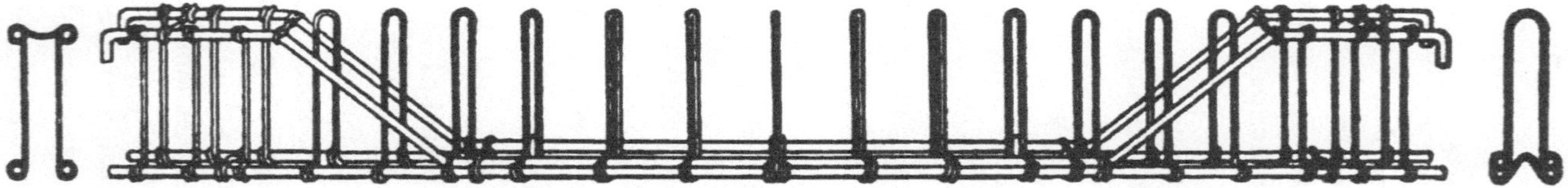

Fig. 72. Unit-Girder Frame System.

Fig. 73. Trussed Bars Used as a Girder Frame.

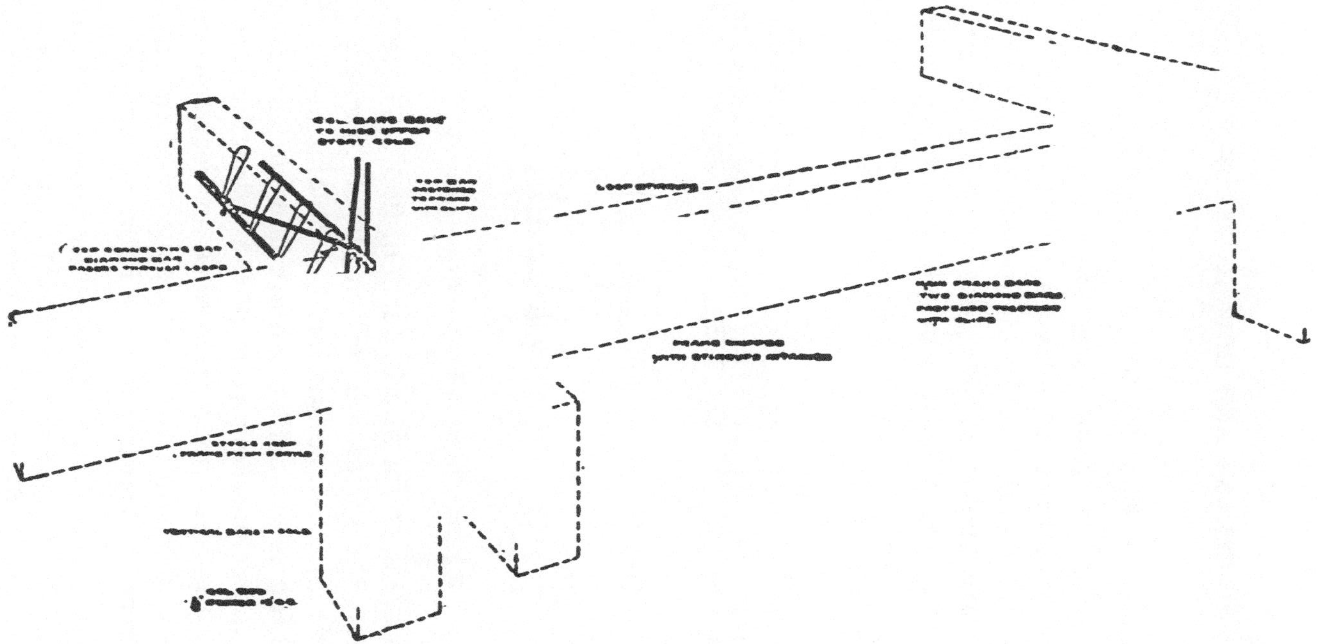

Fig. 74. Steel Concrete Unit-Frame.

Other specimens of this type of reinforcement are shown in Figs. 72, 73, 74, and 75.

Figs. 76 and 77 show two systems of beam and girder reinforcement which are often used.

Several methods used in the reinforcement of floors are shown in the accompanying figures. Fig. 78 shows a system in which rods and wire

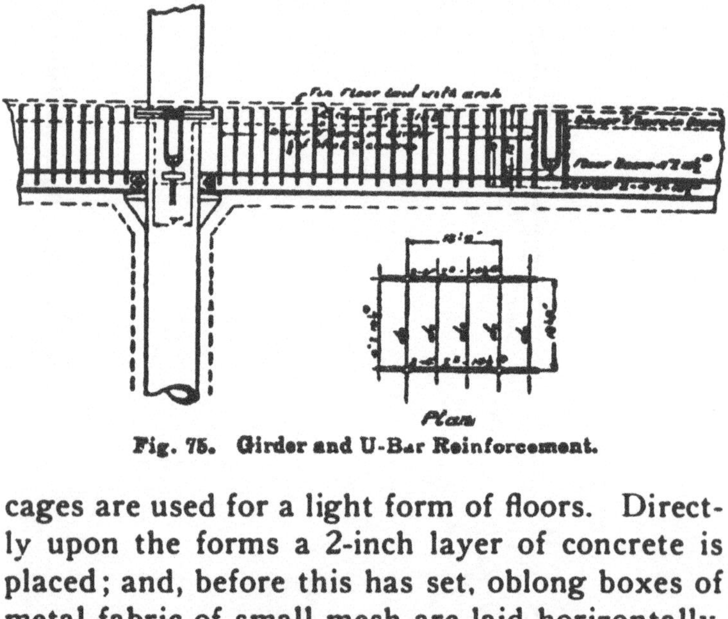

Fig. 75. Girder and U-Bar Reinforcement.

cages are used for a light form of floors. Directly upon the forms a 2-inch layer of concrete is placed; and, before this has set, oblong boxes of metal fabric of small mesh are laid horizontally, with the reinforcing rods in the spaces between them, and the concrete is filled in between the boxes and around the reinforcing rods, and covered over the top to form the floor.

Fig. 79 shows a form of heavier floor construction. The reinforcement in this system consists of flat bars placed upon edge, secured at the ends to the steel beams and bridged with bar separators. The object of the edgewise position of the bars is the increased protection thus secured to the reinforcing steel. With this type of floor the structural steel frame is generally completely encased with concrete.

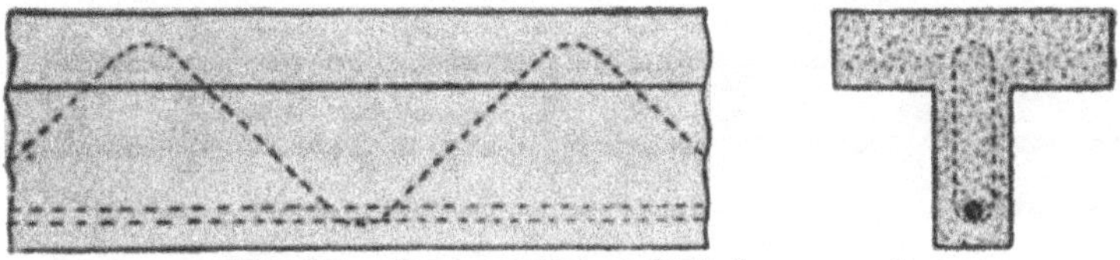

Fig. 76. Bar and Spiral Reinforcement.

For light roof construction where the steel work need not be protected, a continuous slab is built over the beams, reinforced with flat steel bars, $3\frac{1}{16}$ in. by $1\frac{1}{4}$ in., placed edgewise and held in position by spacers as shown in Fig. 79.

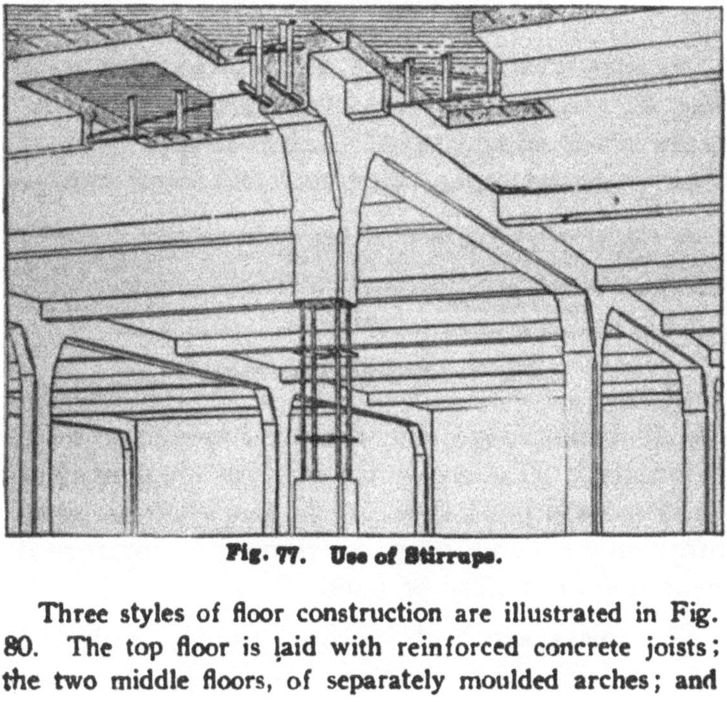

Fig. 77. Use of Stirrups.

Three styles of floor construction are illustrated in Fig. 80. The top floor is laid with reinforced concrete joists; the two middle floors, of separately moulded arches; and the bottom floor of cast slabs, with reinforced ribs moulded on the bottom surface. The thin slabs are also well adapted to roof construction.

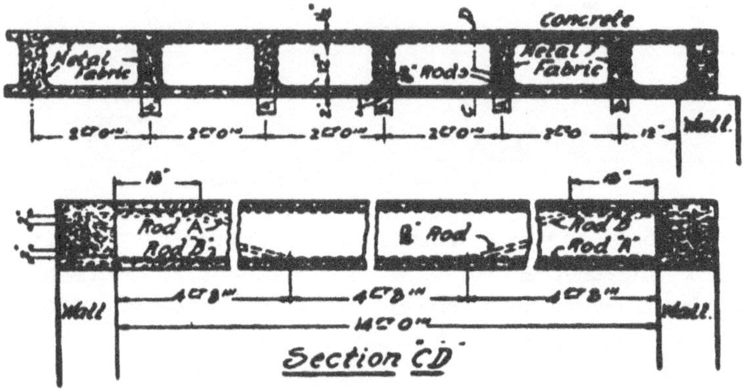

Fig. 78. Light Floor System.

Another system of floor construction, shown in Fig. 82, consists of reinforced concrete "joists" made in an exaggerated I-beam shape. For ordinary construction, the upper and lower flanges

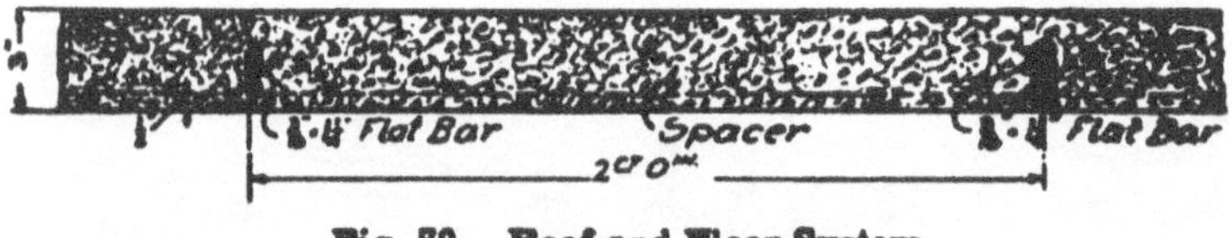

Fig. 79. Roof and Floor System.

are 12 inches wide, and the total height of joists is 8 inches. The upper flange is made somewhat thicker than the lower. Any good system of reinforcement combining tension and horizontal shear members may be used.

The joists are laid in the building with the flanges touching, forming a continuous and com-

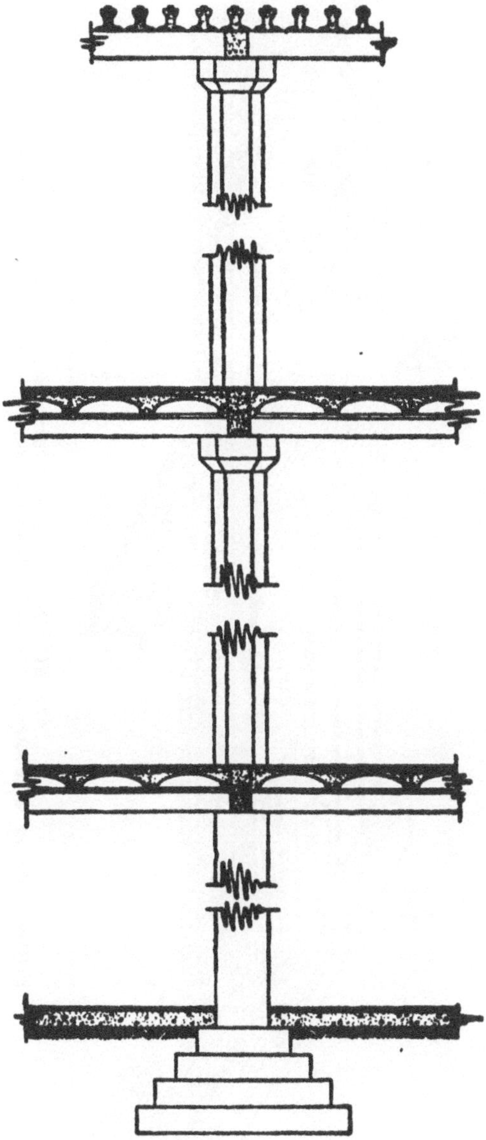

Fig. 30. Three Styles of Floor Construction.

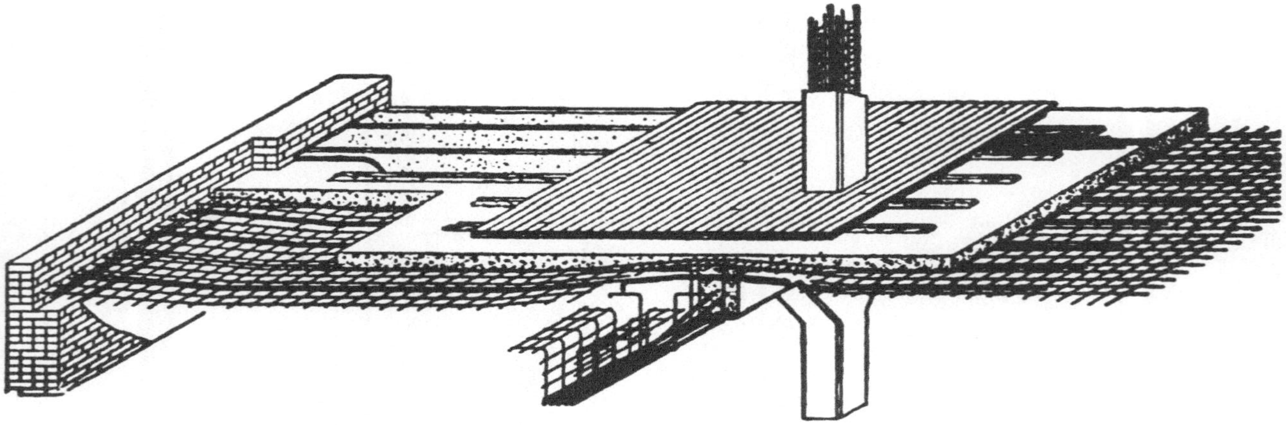

Fig. 81. System of Reinforcement.
Showing practical application of rods and fabrics to girders, columns, and floor and roof slabs.

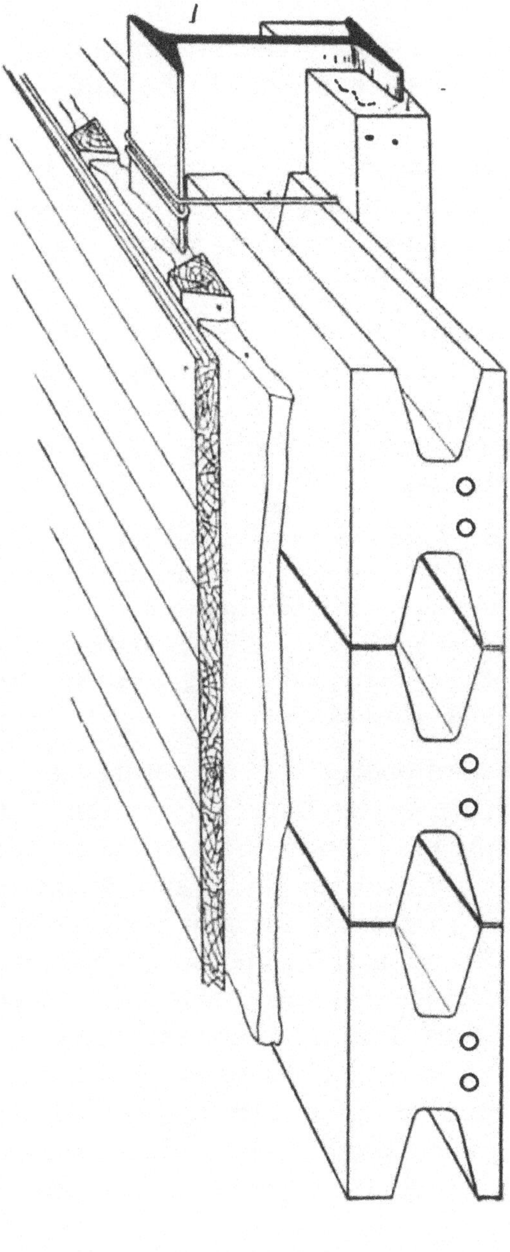

Fig. 82. Concrete Joist Floor.

plete surface on top to hold the finishing floor. Any type of floor—plain cement finish, tile, granolithic, or wood—can readily be laid on the joists.

The under side of the joists forms another complete and continuous surface for the ceiling, requiring but a single coat of rough plaster on which to place the final finish.

Reinforcement for Footings and Foundations. The methods for reinforcing footings and foundations for buildings resting upon soft or unstable ground may be considered as coming under two classes:

1. When the footing is to be spread out wide enough so that the allowable pressure per square foot of bearing area is a safe amount for that particular soil;

2. When concrete piles are used to obtain the same result with the use of less horizontal bearing area and a corresponding lessening of materials used.

Since the load upon a spread footing commonly comes on its center, especially in the case of columns, some means must be used to distribute that load evenly over the under side of the footing, and also to provide for any bending action which may occur in the wide flat surface on account of the concentrated central load. A common way of providing for these necessary points is by the use of rods, bars, I-beams, T-beams, old steel rails, etc., laid in a form of grill work and embedded in concrete. Sometimes they are simply laid at right angles to each other as

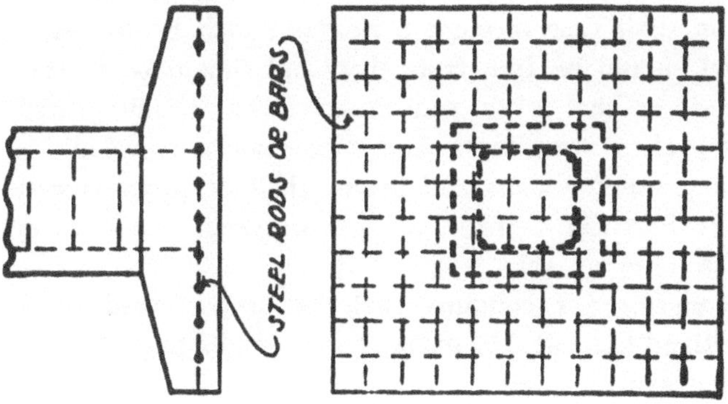

Fig. 83. Reinforced Concrete Column Footing.

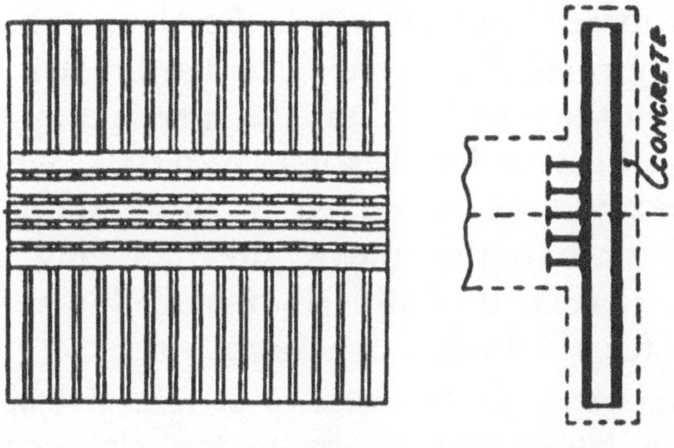

Fig. 84. Reinforced Concrete Footing of I-Beams.

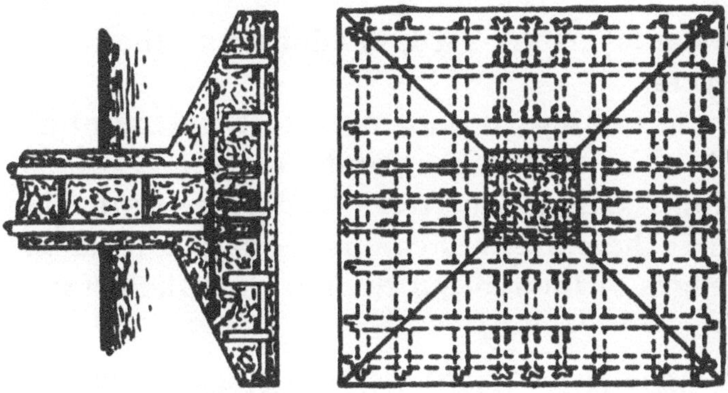

Fig. 85. Column Footing with Stirrups.

shown in Fig. 83 and Fig. 84, and at other times diagonal layers are used, in addition to the former layers.

These footings do not need to be made of an equal thickness throughout, as the greatest bending action will tend to come near the middle; therefore it is common practice to make the edges of the footing thinner than the central part.

SPECIFICATIONS FOR REINFORCED CONCRETE

In reinforced concrete work, Portland cement is the only cement allowed to be used. It must conform to the standard tests as specified by the American Society for Testing Materials, which have already been described in this volume.

In addition to the above, the following are typical requirements ordinarily included in specifications for reinforced concrete:

The aggregate shall be composed of broken stone or gravel. The broken stone shall be of a hard, close-grained quality, free from dust, and crushed so that its largest dimension shall pass through a ring one inch in diameter.

Gravel should be free from dirt, and should be in size ranging from that of a pea to an inch. Disintegrated stone or broken stone containing mica shall be rejected.

All concrete for slabs and beams shall be proportioned of 1 part Portland Cement, 2 parts of sand, and 4 parts broken stone or gravel.

All concrete for columns shall be proportioned of 1 part Portland Cement, 1½ parts of sand, and 2 parts broken stone or gravel.

All concrete shall be machine mixed, using either a batch or a continuous mixer of approved design. Plenty of clean water should be used, so that the resultant mixture has the consistency of what may be known as a "wet" mixture. All materials shall be first thoroughly mixed dry, after which the proper amount of water is added, and the mixing continued until the concrete is uniform. A competent foreman must be in constant attendance at the mixer, to give his approval of every batch which leaves the machine.

No reinforcing steel shall be considered that does not provide for shearing stresses as well as direct tension. Shear members shall be rigidly attached to the main tension member. Sufficient steel shall be placed so that concrete will be obliged to resist only direct compression, and shearing stress up to 50 pounds per square inch. No steel shall have, at any point, less than 1 inch of concrete covering. Steel bars shall not be painted. A slight film of rust will not be objectionable upon same, but any bar on which decided rust scales have formed shall be rejected. In no case shall steel of higher elastic limit than 45,000 lbs. be considered. The same shall have a tensile strength of from 60,000 to 70,000 lbs. per square inch, with an elongation of not less than 20 per cent in 8 inches. A bar should bend when cold, around its own diameter, through an angle of 180 degrees, and close down upon itself without cracking.

Steel shall be placed in exact accordance with detail drawings.

Concrete shall be placed as rapidly as possible after mixing, and shall be thoroughly puddled immediately thereafter.

When concreting is once commenced, it should be carried on vigorously to completion, if possible. If concreting must be stopped before an entire floor is completed, the stop shall be made in the center of beams and center of floor slabs.

Concrete shall be placed in freezing weather, only when same cannot possibly be prevented: and then especial precautions must be used to cover the work at once with at least five or six inches of sawdust or manure. The forms for such work should be left in place at least three weeks longer than customary. Concrete for slabs shall be laid immediately after pouring the beams.

The centering must be of sufficient strength to carry easily the dead weight of the construction as a liquid, without deflection.

The design of the centering should be such that the sides of the beams can be taken down first, then the slab centering complete. Centering should never be removed until the concrete has thoroughly set, and has aged to give it sufficient strength to carry its own weight besides whatever live load is liable to come on the work during the course of construction. In no event shall the falsework be removed until approval is given by the architect or engineer-in-charge. Beams shall remain supported for at least two weeks after all other falsework has been removed. Columns shall not be given their full loading in less than five weeks after concreting.

Floors shall be tested after the centering has been removed one month, to a uniformly distributed load equal to twice the safe live load. With this load there should not be a deflection exceeding 1-400th part of the span, and the floor should return to its normal position after the removal of the load.

The following stresses based on figuring full live and dead loads shall be used in the design:

For hooped concrete columns, 750 lbs. per sq. in. of concrete.

For latticed columns, 500 lbs. per sq. in.

For shearing stresses in concrete and adhesion of concrete to steel, 50 lbs. per sq. in.

Extreme fiber stress in compression for slab, beams, and girders, 750 lbs. per sq, in.

Tensile stress in steel, 16,000 lbs. per sq. in.

The ratio of moduli of elasticity of concrete and steel to be 1 to 15.

The tensile strength of concrete shall not be considered.

When slab and beams are built continuous over their supports, the bending moment may be taken at 1-10 Wl. In cases of square floor panels, floor supported on all four sides and built continuous over supports, the bending moment may be taken at 1-20 Wl. In all such cases sufficient reinforcement must be provided at the top of the slab to take care of the regular bending moment at the supports.

FIREPROOF CONSTRUCTION FLOORS, SLABS, AND ROOFS

A reinforced concrete slab is made by placing in equally spaced parallel rows some one of the materials used for reinforcement and covering same with concrete.. If these rows run only one way, the construction is called an independent bar reinforcement. If they cross transversely, the result is a latticed reinforcement. Latticed reinforcement is well adapted to floor and roof work as the transverse members of the reinforcement prevent shrinkage cracks.

Reinforced concrete floor construction may again be divided into three classes:

(1) Those constructions which serve simply as a filling between the girders and beams of a floor framework of steel;

(2) Those in which the girders and beams are themselves reinforced concrete;

(3) Those in which the girders and beams are done away with.

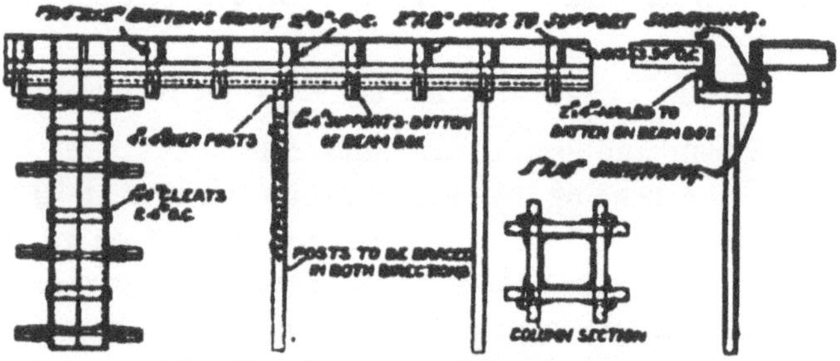

Fig. 86. Beam and Column Forms.

The first class may be divided again into flat slab and arched slab constructions. Fig. 87 shows a type of arched floor. This type of floor is commonly used to support heavy loads.

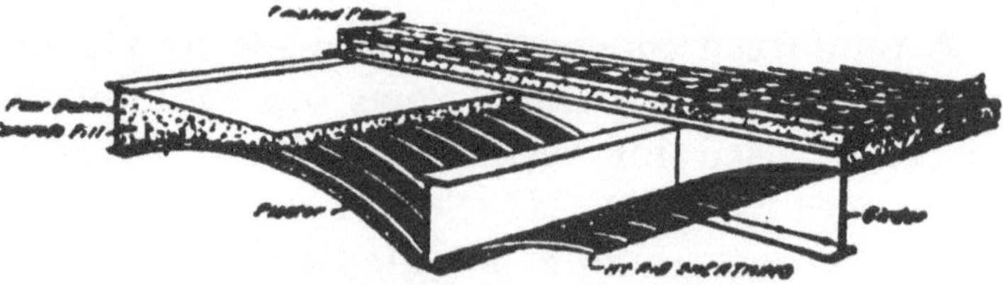

Fig 87. Arched Floor Construction.

In these structures of the first class, the floor-slab either rests on top of the beam or girder, embeds the top flange, or rests upon the bottom flange.

The second class comprises floors of such a construction that the girders and beams really constitute ribs for strengthening the slabs. These are monolithic constructions.

The third class, comprising those monolithic structures in which the beams and girders do not project, is illustrated by the "Mushroom" system, the "Spider-Web" system, or the "Umbrella" system.

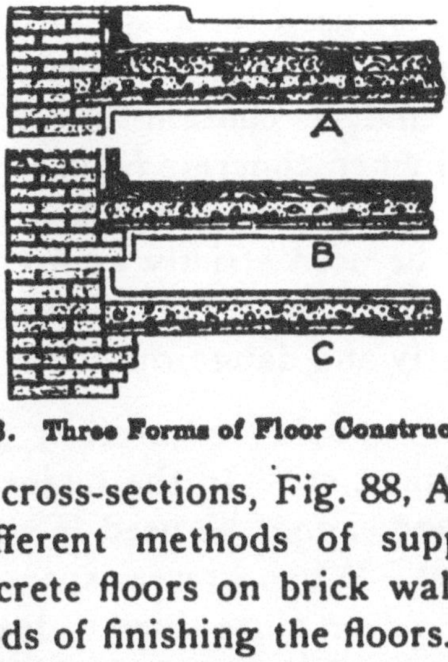

Fig. 88. Three Forms of Floor Construction.

The three cross-sections, Fig. 88, **A**, **B**, and **C**, illustrate different methods of supporting reinforced concrete floors on brick walls, and different methods of finishing the floors. In **A** and **B**, the plastering is applied directly to the under side of the floor slabs. In **B** an under-floor is first nailed to the concrete, and to this under-floor the finished floor is nailed.

Fig. 88, **C**, represents a plain reinforced floor with a surface finish of a richer grade of concrete. The ceiling below is formed by the cleaned under side of the floor above.

In each of these illustrations, the reinforcing agent is **expanded metal.**

Cinder Concrete. On account of the lightness of cinders and the peculiar elasticity of the concrete made from them, to expand and contract in resisting heat, **cinder-concrete fireproofing** has largely been used for floor-slabs between steel beams.

Cinders make good fireproof concrete and this material may be used for suspended ceilings and as a filler on top of slabs between furring strips. Sometimes cinders contain unconsumed coal, which is bad when concrete is mixed wet, and is a dangerous element in case of fire. Cinder concrete should be used strictly as fireproofing and not as a structural material. Used in the latter way, it is costly and dangerous.

The cinders should be carefully selected and freed from ashes, etc. In the majority of cities, cinder concrete cannot be used in spans exceeding 4 to 6 feet. When greater spans are permitted, they are never more than 10 feet, and must be designed after testing a sample slab or beam to destruction, and using data thus obtained with a factor of safety of ten. The ultimate strength of good, well-made cinder concrete is about 900 pounds per square inch in compression at the end of thirty days with a 1:2:4 mixture. It is now recognized as one of the speediest and most convenient forms of floor construction.

CEILINGS

Ceilings may be formed in any of the following ways, depending upon the character and construction of the building:

By plastering directly upon the under side of floors, which are commonly of sufficient roughness from the forms so that they will hold the plaster.

By plastering on a surface of metal lath suspended from steel bars which are fastened to the steel work of the building frame.

By smoothing up the bottom surfaces of concrete joists, when used side by side as a floor, applying a single coat of plaster, and then a finishing coat.

By the use of carefully jointed and smoothed floor forms, upon which the concrete is carefully spaded and worked over in placing. This allows the mortar to sink down upon the forms and furnishes a smooth ceiling for the room below.

ROOFS

The principles governing the design of reinforced concrete roofs are similar to those for floors. The reinforcing materials generally

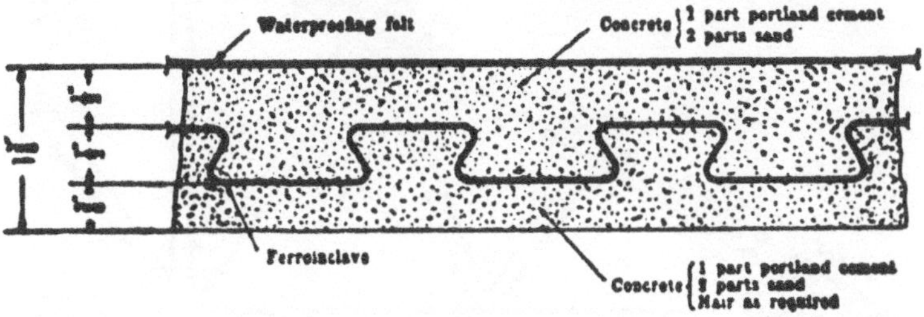

Fig. 89. Section of Concrete Roof.

used for roofs are of light weight. The general construction is one of a combination of light

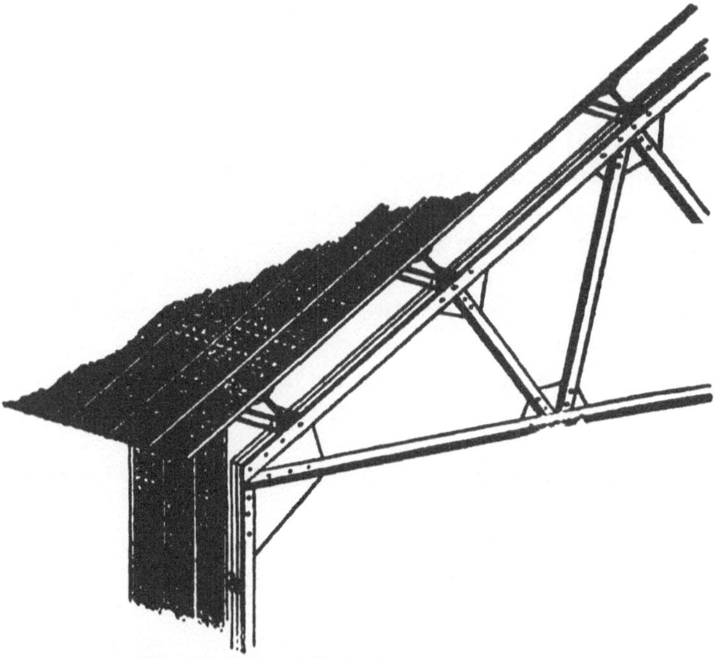

Fig. 90. Metal Lath Roof Construction.

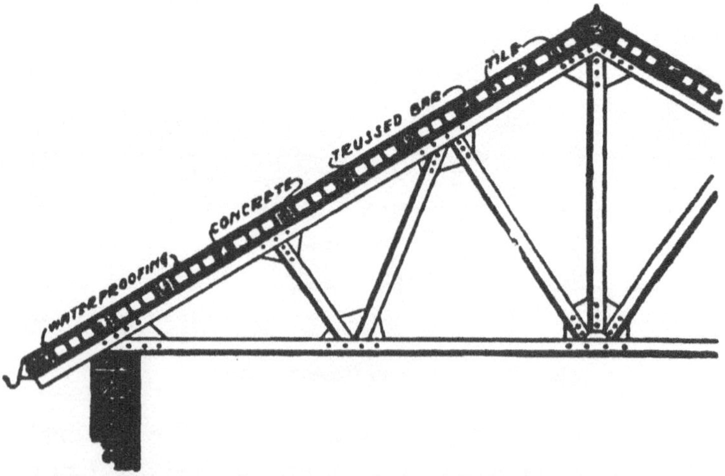

Fig. 91. Tile and Reinforced Concrete Roof.

steel shapes for the framework, upon which is laid some form of sheet fabric such as expanded metal, rib metal, closely woven wire fabrics, etc. The closer woven metals, when used, do away with the use of wooden forms, thus allowing the

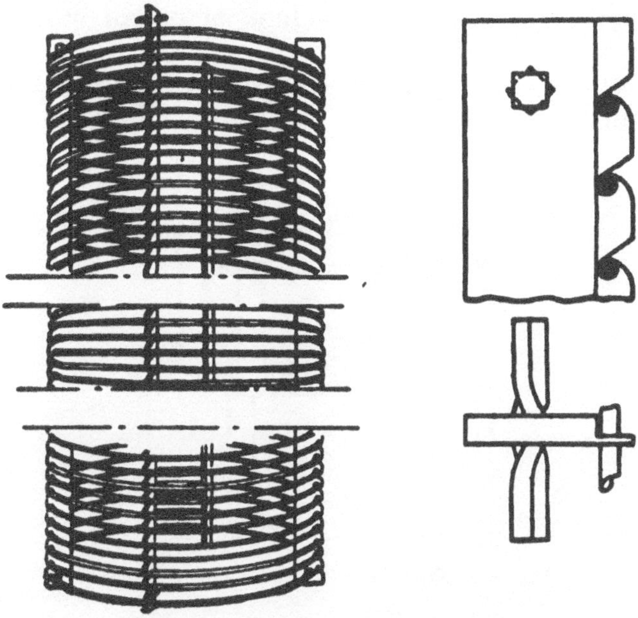

Fig. 92. Details of Spiral Column.

concrete to be deposited upon the metal directly; the holes in the fabric letting enough of the wet mixture pass through to form a good bond, and also to provide a rough surface to hold the plaster for finish on the under side.

COLUMNS

The general plan of all reinforced columns seems to be that of a cage supported by upright members and filled with concrete, the outside being likewise protected from fire and corrosion by a thick layer of concrete. This cage may be made up in various ways. Two of the common ways are as follows.

(a) By the use of some form of coil of metal surrounding the vertical bars;

(b) By the use of tie-rods spaced regularly up the length of the column,

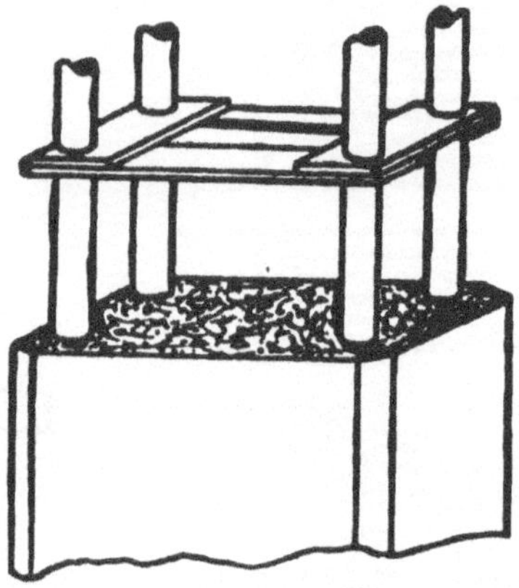

Fig. 93. Column with Rod Reinforcement.

Figs. 92 and 93 show the scheme of the two methods.

WALLS

In small structures, it is common practice to build the walls as one continuous reinforced piece, hollow or solid as desired, the forms being carried up as the wall progresses. The reinforcement in such walls consists of lateral rods, wires, expanded metal, or wire fabric, as thought best in the individual case. Window- and door-frames should be thoroughly reinforced to prevent the formation of cracks from one to another.

Metal ties used in hollow walls should have their surfaces covered with cement to prevent corrosion and gradual wasting away. At the corners of the building, the reinforcing rods or material used, from the two sides, should lap over each other so as to make a firm corner joint and tie the two walls together. This applies to both double and solid walls.

In the monolithic form of reinforced concrete building, the space between the outside beams, girders, and columns on the ends and sides may be filled in with a curtain wall of brick, tile, or even a reinforced concrete slab with bars running two ways for strength and to prevent shrinkage. In residence work, walls are sometimes built double, with a 4-inch air-space between the two reinforced slabs. In this type of wall, the reinforcing rods should also run horizontally and vertically.

The thickness of slab and amount of reinforcement necessary for a **curtain wall**, or a vertical wall which is to bear no weight, is determined by figuring it as a flat slab supported at all four sides, and carrying a uniformly distributed load of 40 lbs. per square foot due to wind pressure. An ordinary slab designed on this basis will probably be four or five inches thick, allowing for a good factor of safety.

PARTITIONS

Partitions in buildings of heavy construction, or where any considerable load is to be carried, are generally made solid and varying from 3 to 6 inches in thickness. These partitions may be

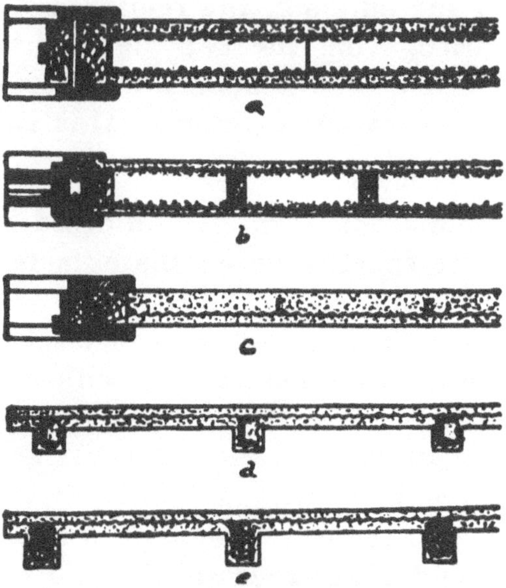

Fig. 94. Types of Partitions.

poured by the use of vertical forms as in the construction of walls. Often slots are left in concrete floors so that the partitions may be poured after the floors are laid.

Hollow partitions are studded, and metal lathing of some form is held by metal fastenings on each side of the studding. A heavy coat of mortar is then plastered on each side of the metal wall thus formed.

If expanded metal is used as a reinforcing agent in partitions, always lay the metal with the length of the diamond across the shortest span; it has only half the strength when placed the other way.

CONCRETE BRIDGES AND CULVERTS

From the point of view of strength and stability, it is easy to show the advantages of reinforced concrete for bridge construction. Reinforced concrete is now well understood, and designs may be made upon a rational basis, the steel carrying the tensile stresses, and the concrete the compressive stresses. The steel is well

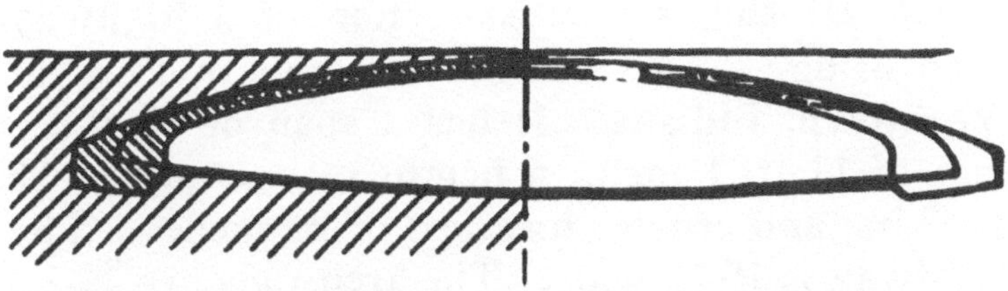

Fig. 95. Cross-Section of Highway Arch Bridge.

protected and can never deteriorate, while the concrete gains in strength as the years go by. The mass and weight of the concrete work make it proof against vibration and the danger of floods and winds.

The first cost of reinforced concrete bridges is slightly greater than that of wooden bridges, about the same as that of steel bridges for short spans, and considerably less than that of stone bridges. The cost of maintenance is a large item for wooden and steel bridges, and both kinds must be renewed within 10 to 25 years; the maintenance cost for a concrete bridge is nothing, and the bridge should last forever.

Highway Arch Bridges. Highway arch bridges are usually low and flat, and vary in span length from 20 to 100 feet. The ordinary type consists of an arch-barrel and spandrel walls, with an earth filling between the walls. Ornamental concrete parapets are often used, although ordinary gaspipe railing is very common. The width of these bridges ranges from 14 feet for small creek crossings, to 60 feet or over for structures carrying city streets

Fig. 95 shows a cross-section of a highway arch bridge. This bridge was built in 1904, at Yorktown, Indiana. It has a span of 95 ft.; a rise of 11 ft. 1 inch.; a height of opening of 15 ft. 7 in., and crown thickness of 26 inches. The roadway is 16 ft. wide. The arch rods are ¾ in. square, spaced 6 inches. The parapets are monolithic concrete, 18 in. by 3 ft. high. The design was made for a 20-ton roller, or 200 pounds per sq. ft.

For ordinary highway construction where long spans are required, and where heavy loads such as traction engines and their accompanying machines or where heavy earth-fills will occur, an arch of this type is especially suitable.

In cities and towns where it is necessary to have the streets cross a stream, the reinforced concrete arch not only insures strength and stability, but is capable of such varied artistic treatment that it is being almost universally adopted in municipal work.

Girder and Slab Bridges. Girder and slab bridges are known as **beam** bridges; and carry the loads which come upon them by their resist-

Fig. 96. Cross-Section of a Concrete Girder Highway Bridge.

Fig. 97. Cross-Section of a Concrete Slab Highway Bridge.

ance to bending. They are extensively used for short spans and low, shallow crossings.

The **flat slab bridge** is the simplest to design and construct, and also proves to be the most economical in materials when used for spans up to about 20 feet. For spans from 20 to 35 feet or thereabouts, the girder type is the best to use. Beams longer than 35 feet are rarely built; an arch is generally used for such spans.

Abutments and Piers. The supports at each end of a bridge are known as **abutments**. An abutment serves a number of purposes: it furnishes a resting-place for the end of the span; it distributes the weight of the bridge over the foundation; and it prevents the bank from running out into the stream at the end of the bridge.

For the average size of bridge and height of abutment, wing-abutments are the most generally used, and are probably the best. Concrete

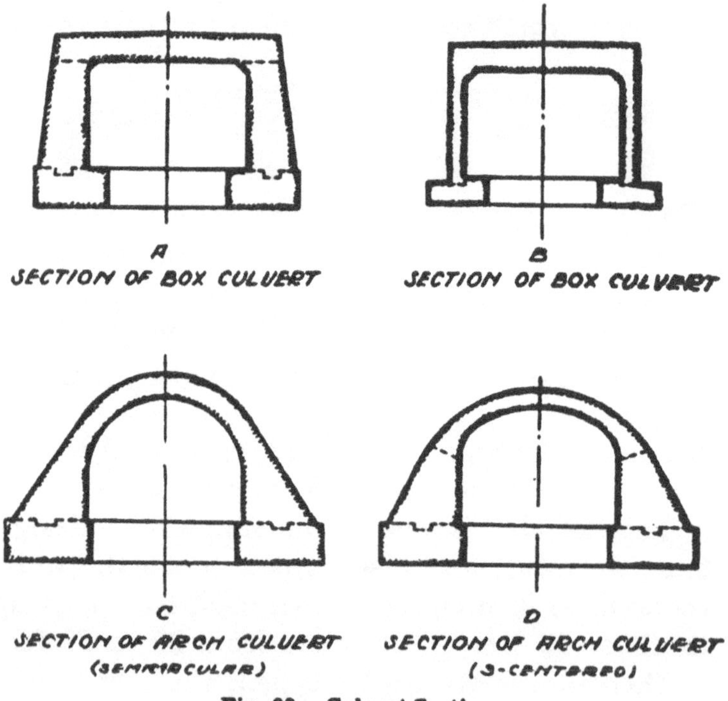

Fig. 98. Culvert Sections.

has replaced stone masonry as the material for constructing such abutments, and answers the purpose excellently. Wing-abutments are so called because of the two sloping walls which extend out from the sides to the full width of the embankment to prevent the earth from running into the stream. These walls are known as **wing-walls.** The entire abutment, including the wing-walls, must be designed to retain the earth behind it.

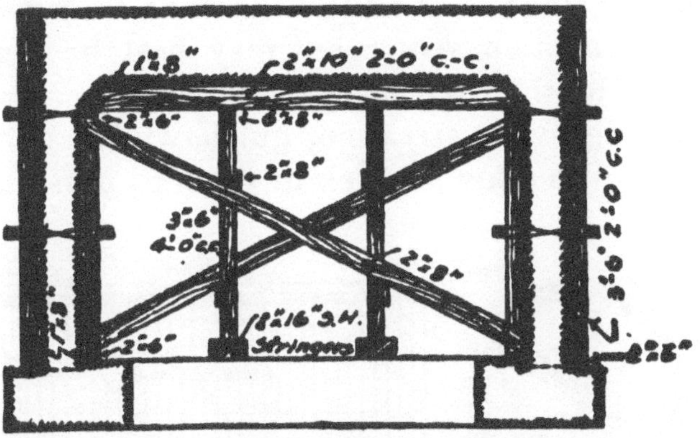

Fig. 99. Section of Concrete Box Culvert, Showing Forms.

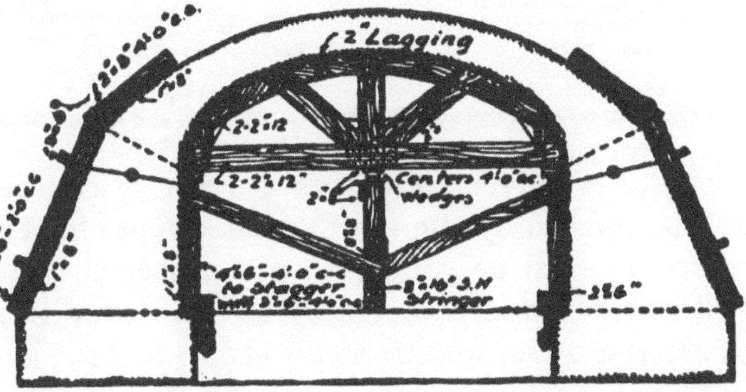

Fig. 100. Cross-Section of Concrete Arch Culvert, Showing Forms.

Bridge piers have been built of piles, timber frames filled with stone, metal cylinders filled with concrete, brick, and stone masonry, but are now generally constructed of concrete, which is an ideal material to answer all the requirements.

Concrete Culverts. The difference between a culvert and a bridge is not very clearly marked; a large culvert may be called a small bridge, and a small bridge is but little different from a culvert. The two structures may best be distinguished by their purpose; a bridge is intended as a crossing over a stream or gulch, while a culvert is needed to allow a creek or ditch to pass under

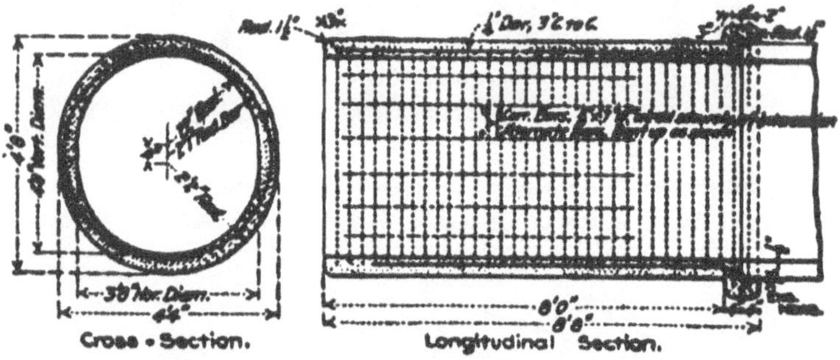

Fig. 101. **Fig. 102.**
Sectional Reinforced Concrete Culvert Pipe with Bell and Spigot Ends.

a road. A culvert serves as a drain or conduit, and is covered by the road embankment; for this reason it must be composed of a material which will not rot under the most adverse conditions of alternate wet and dry periods.

There are three distinct types of culverts: **Box-culverts,** which are rectangular in section; **arch culverts,** which have arched tops and in some cases arched floors; and **pipe culverts,** which are circular or elliptical in section.

Box culverts (see Fig. 98, A and B, and Fig.

99) were formerly built of wood or stone, and later of concrete, with covers of old rails; but reinforced concrete is now the standard construction. Arch culverts (see Fig. 98, C and D, and Fig. 100) are built of concrete, either plain or reinforced, the more modern material having replaced stone masonry.

Culvert pipes of cast iron and clay products are still used, but concrete has invaded this field

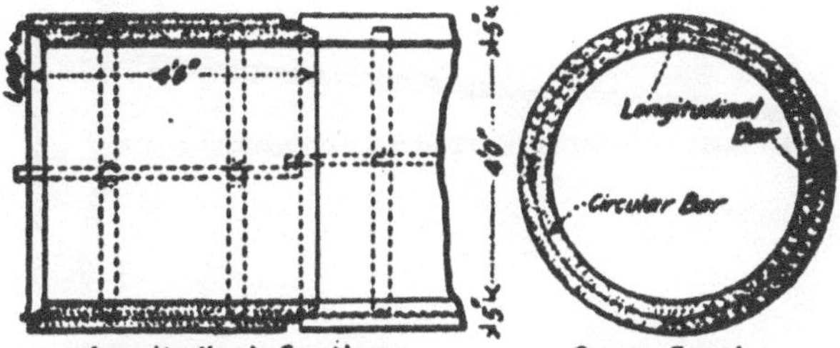

Longitudinal Section. Cross-Section.

Fig. 103. Fig. 104.
Sectional Concrete Pipe with Tapered Ends.

also, and it will probably supersede the other materials.

Reinforced Concrete Culvert Pipe. Figs. 101 to 104 illustrate the sectional concrete culvert pipe now being used in many cases instead of iron pipe. The sections are made 6 or 8 feet long so that they may be easily handled. The bell and spigot ends (Figs. 101 and 102) provide a good and flexible means of connection, but are expensive to construct; and tapered ends, as in Figs. 103 and 104, are more often used.

Concrete Retaining Walls. A retaining wall

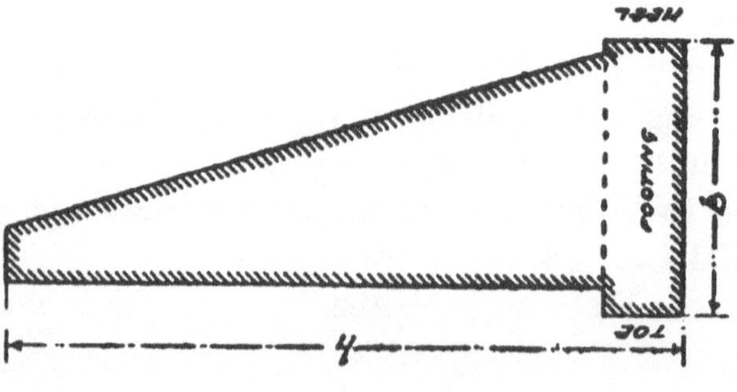

Fig. 105. Section of Gravity Retaining Wall.

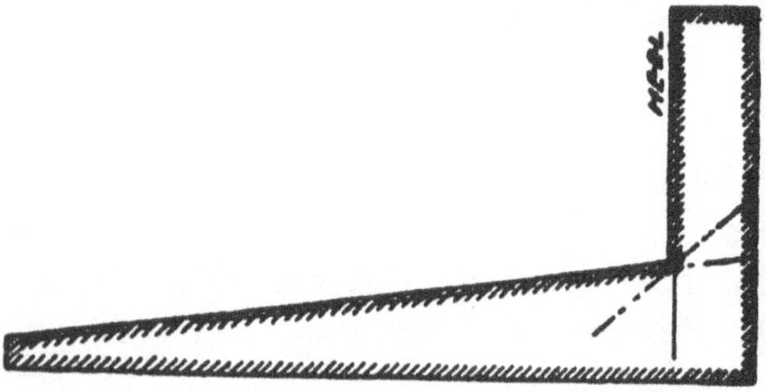

Fig. 106. Retaining Wall of L-Section.

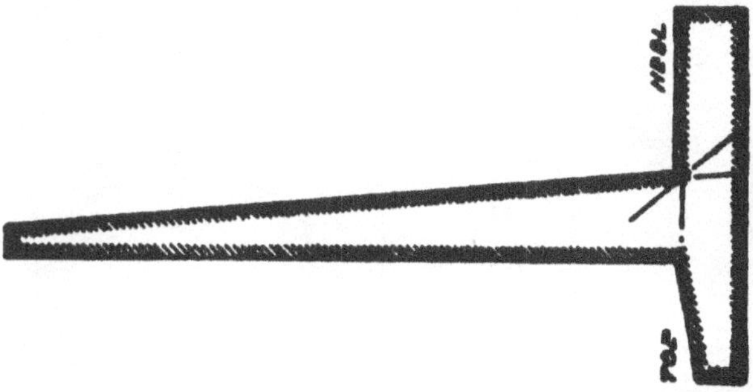

Fig. 107. Retaining Wall of T-Section.

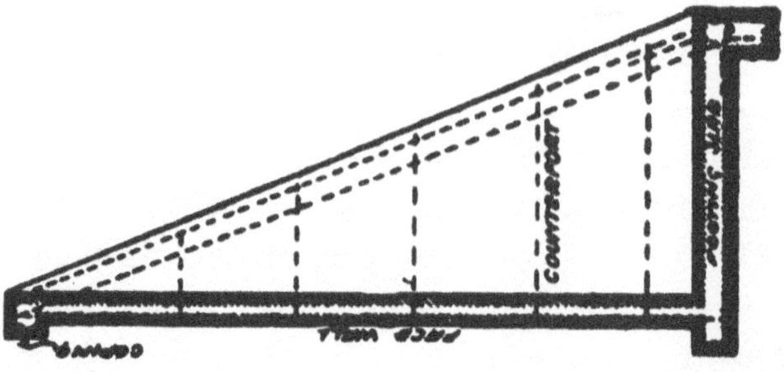

Fig. 108. Section of Counterfort Retaining Wall.

is a wall built for the purpose of supporting or holding back a bank of earth, rock, or other loose material. Such a wall is required where-ever it is necessary to have an abrupt change in the level of the ground surface, and where it is not desirable, because of lack of space or for some other reason, to let the earth run out to its natural slope.

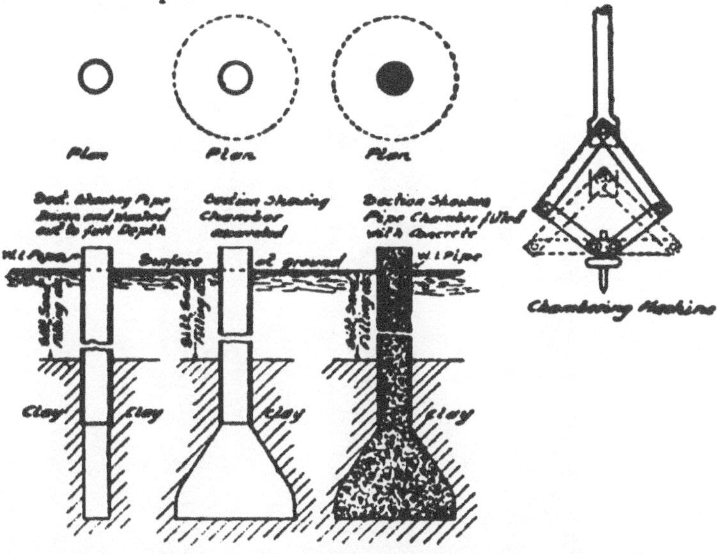

Fig. 109. One Method of Poured Concrete Piling with Extended Footing.

Because retaining walls are practically buried in the ground, and because they must endure in-definitely, they are constructed of masonry. Stone masonry walls depend on their weight to hold back the earth behind them and are sel-dom built nowadays, concrete being used instead. The use of reinforced concrete has made it pos-sible to construct a number of economical types

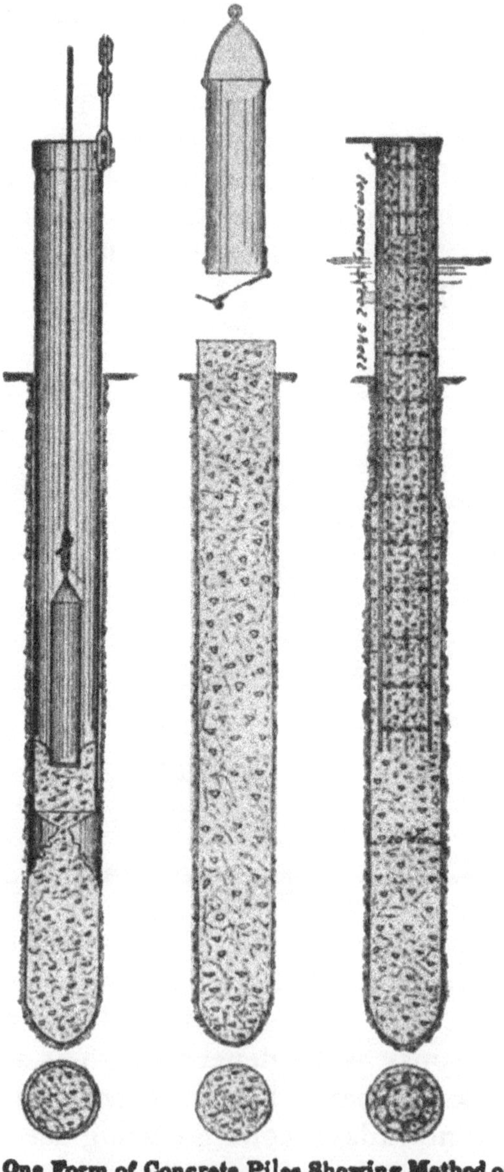

Fig. 110. One Form of Concrete Piles Showing Method of Placing Concrete and Reinforcement.

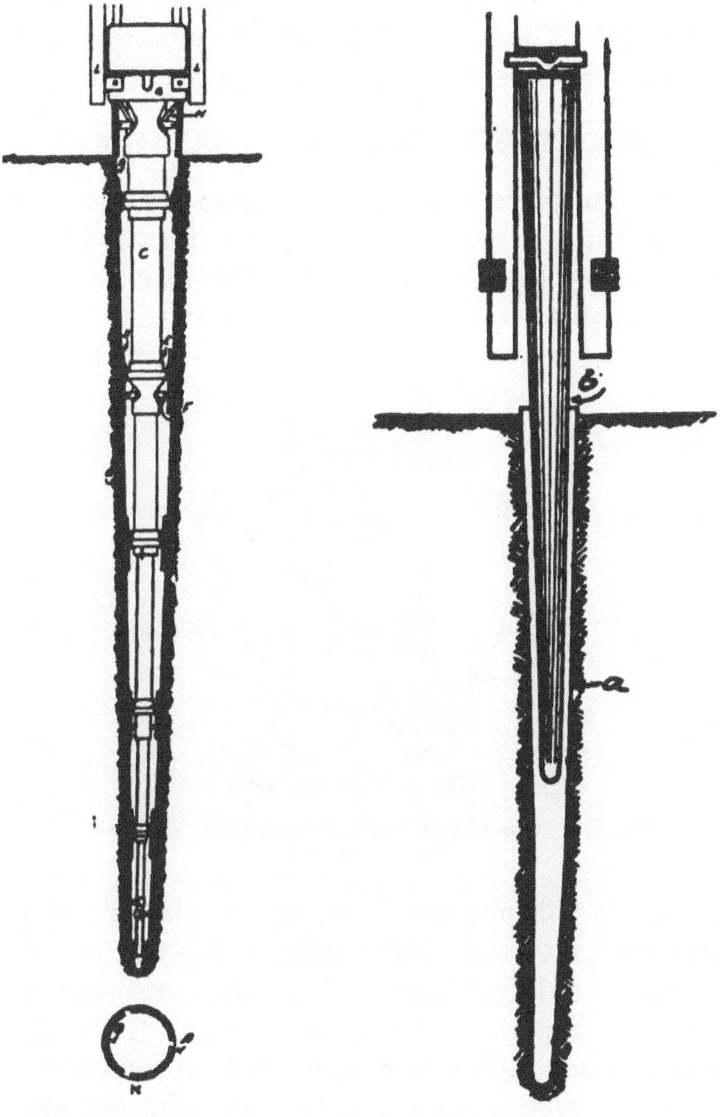

Fig. 111. Shell and Core Type of Concrete Pile.
Showing Section of Pile Core. **Showing Driven Shell.**

of wall which could not be built of plain concrete or stone. The common types of retaining wall section are illustrated in Figs. 105 to 108.

CONCRETE PILE FOUNDATIONS

The use of reinforced concrete in foundations in the form of concrete piles, is a factor which promises to be of great value in the future. The one advantage over concrete piling that wood piling has possessed—namely, low initial cost—

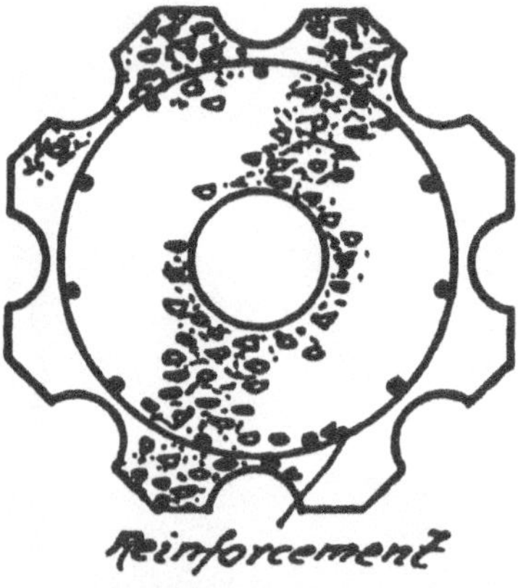

Fig. 112. Cross-Section of Corrugated Pile Showing Reinforcement.

is rapidly disappearing, because of the growing scarcity of the available lumber supply due to constant deforestation. The absolute permanence of concrete piling, its freedom from the dangers that threaten the integrity of wood piling—rot, over-driving, the attacks of boring animals, etc.,— its low ultimate cost, and the fact that its constituent materials may be obtained almost anywhere, are factors that in time will drive wood piling out of general practice.

There are two general methods of concrete pile construction: those constructed in place, and those moulded or rolled in advance and driven by methods similar to those used for driving wooden piles. In the first method some form of collapsible steel core, encased in a closely fitting

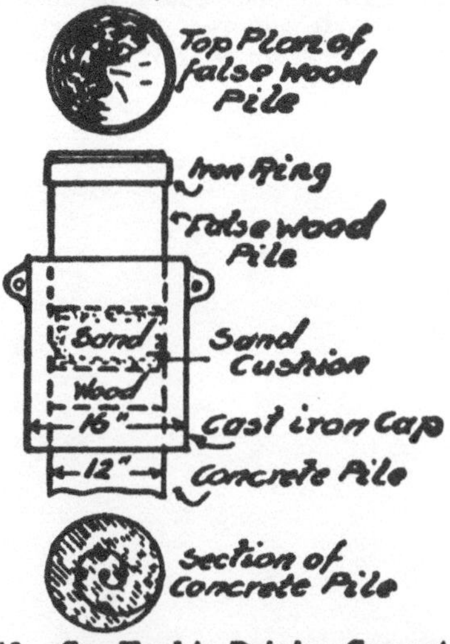

Fig. 113. Cap Used in Driving Concrete Pile.
Section of Rolled Pile with Reinforcement is Shown at Bottom.

shell of suitable material, is driven in the usual manner, the core withdrawn and the shell filled with concrete. Piles constructed in this manner may be either plain or reinforced, depending largely upon the character of the work for which they are to be used. In the second method it is quite essential that they be reinforced to withstand the strains to which they are subjected in handling and driving.

In certain cases concrete piles are an economical substitute for deep pier foundations.

ACKNOWLEDGMENT.

In the compilation of such a work as this, it is necessary to draw to some extent not only from the standard literature, but also from the current trade literature of the subject. In this connection, the Editors wish to express their thanks to the Atlas Portland Cement Company, the Universal Portland Cement Company, and various other manufacturers and dealers in the materials and machinery of construction, also to the Association of American Portland Cement Manufacturers, and to proprietors and patentees of special machinery and processes, for many helpful suggestions and for valuable data and illustrations, which are here reproduced by special permission.

INDEX

Spotted something that isn't quite right?

Let us know and we'll fix it for future reprints.
And as a thank you for your diligence, we'll
send you a free Sothis book of your choice.

publisher@sothis.press

This page intentionally left blank

Made in United States
Cleveland, OH
06 November 2026

47791723R00227